WHAT EVOLUTION IS

Also by Ernst Mayr

This Is Biology (1997)

One Long Argument (1991)

Principles of Systematic Zoology (1990, with Peter Ashlock)

Toward a New Philosophy of Biology (1988)

The Growth of Biological Thought (1982)

The Evolutionary Synthesis (1980, with W. Provine)

Evolution and the Diversity of Life (1976)

Populations, Species, and Evolution (1970)

Principles of Systematic Zoology (1969)

Animal Species and Evolution (1963)

Methods and Principles of Systematic Zoology (1953, with E. G. Linsley and R. L. Usinger)

Birds of the Philippines (1946, with Jean Delacour)

Birds of the Southwest Pacific (1945)

Systematics and the Origin of Species (1942)

WHAT
EVOLUTION
IS

ERNST MAYR

BASIC
BOOKS

A Member of the
Perseus Books Group

To the naturalists from Aristotle to the present,
who have taught us so much about the living world.

Copyright © 2001 by Ernst Mayr

Published by Basic Books,
A Member of the Perseus Books Group

A CIP catalog record for this book is available from the Library of Congress.
ISBN-10 0-465-04425-5 (hc)
ISBN-13 978-0-465-04425-2 (hc)
ISBN-10 0-465-04426-3 (pbk)
ISBN-13 978-0-465-04426-9 (pbk)

EBA   06 07 08   20 19 18 17 16 15 14 13 12 11

CONTENTS

Evolution is the most profound and powerful idea to have been conceived in the last two centuries. It was first developed in detail with the 1859 publication of the book *On the Origin of Species*, by Charles Darwin, who enjoyed a long and incredibly productive life. While Darwin's professional career began with a round-the-world biological collecting trip on which he embarked at the age of 22 aboard the HMS *Beagle*, he had already been devoted to outdoor natural history as a boy.

A great deal new has been learned about the workings of evolution since Darwin's day. Wouldn't it be wonderful if Darwin himself, a clear and forceful writer as well as the greatest biologist of his generation, could write for us a new book on the status of evolutionary thought today! Of course that's impossible, because Darwin died in 1882. This book is the next best: it has been written by a man who is one of the greatest biologists of our own day, who has also enjoyed a long and incredibly productive life, and who is also a clear and forceful writer.

To place Ernst Mayr in perspective, I'll relate an experience of my own. In 1990 I carried out the second bird survey of the Cyclops Mountains, a steep, high, isolated range rising from the north coast of the tropical island of New Guinea. The survey proved to be difficult and dangerous, because of the daily risks of falling off the steep slippery trails, of getting lost in the dense jungle, of exposure in cold wet conditions, and of potential conflicts with local people on whom I depended but who had their own agendas. Fortunately, New Guinea had by then been "pacified" for many years. Local tribes were no longer at war with each other, and European visitors were a familiar sight and were no longer at risk of being murdered. None of those advantages existed in 1928, when the first bird survey of the Cyclops Mountains was carried out. I found it hard to imagine how anyone could have survived the difficulties of that first survey of 1928, considering the already-severe difficulties of my second survey in 1990.

That 1928 survey was carried out by the then–23-year-old Ernst Mayr, who had just pulled off the remarkable achievement of completing his Ph.D. thesis in zoology while simultaneously completing his pre-clinical studies at medical school. Like Darwin, Ernst had been passionately devoted to outdoor natural history as a boy, and he had thereby come to the attention of Erwin Stresemann, a famous ornithologist at Berlin's Zoological Museum. In 1928 Stresemann, together with ornithologists at the American Museum of Natural History in New York and at Lord Rothschild's Museum near London, came up with a bold scheme to "clean up" the outstanding remaining ornithological mysteries of New Guinea, by tracking down all of the perplexing birds of paradise known only from specimens collected by natives and not yet traced to their home grounds by European collectors. Ernst, who had never been outside Europe, was the person selected for this daunting research program.

Ernst's "clean-up" consisted of thorough bird surveys of New Guinea's five most important north coastal mountains, a task whose difficulties are impossible to conceive today in these days when bird explorers and their field assistants are at least not at acute risk of being ambushed by the natives. Ernst managed to befriend the local tribes, was officially but incorrectly reported to have been killed by them, survived severe attacks of malaria and dengue and dysentery and other tropical diseases plus a forced descent down a waterfall and a near-drowning in an overturned canoe, succeeded in reaching the summits of all five mountains, and amassed large collections of birds with many new species and subspecies. Despite the thoroughness of his collections, they proved to contain not a single one of the mysterious "missing" birds of paradise. That astonishing negative discovery provided Stresemann with the decisive clue to the mystery's solution: all of those missing birds were hybrids between known species of birds of paradise, hence their rarity.

From New Guinea, Ernst went on to the Solomon Islands in the Southwest Pacific, where as a member of the Whitney South Sea Expedition he participated in bird surveys of several islands, including the notorious Malaita (even more dangerous in those days than was New Guinea). A telegram then invited him to come in 1930 to the American Museum of Natural History in New York to identify the tens of thousands of bird specimens collected by the Whitney Expedition on dozens of Pacific Islands. Just as Darwin's "explorations," sitting at home, of collections of barnacles were as important to Darwin in forming his insights as was his visit to the Galapagos Islands, so too Ernst Mayr's "explorations" of bird specimens in museums were as important as his fieldwork in New Guinea and the Solomons in forming his own insights into geographic variation and evolution. In 1953 Ernst moved from New York to Harvard University's Museum of Comparative Zoology, where even today he continues to work at the age of 97, still writing a new book every year or two. For scholars studying evolution and the history and philosophy of biology, Ernst's hundreds of technical articles and dozens of technical books have been for a long time the standard reference works.

But in addition to gaining insights from his own fieldwork in the Pacific and from his own studies of museum bird specimens, Ernst has collaborated with many other scientists to extract insights from other species, ranging from flies and flowering plants to snails and people. One of those collaborations transformed my own life, just as the meeting with Erwin Stresemann transformed Ernst's life. While I was a teenaged schoolboy, my father, a physician studying human blood groups, collaborated with Ernst in the first study proving that human blood groups evolve subject to natural selection. I thereby met Ernst at dinner at my parents' house, was later instructed by him in the identification of Pacific island birds, began in 1964 the first of 19 ornithological expeditions of my own to New Guinea and the Solomons, and in 1971 began to collaborate with Ernst on a massive book about Solomon and Bismarck birds that we completed only this year, after 30 years of work. My career, like that of so many other scientists today, thus exemplifies how Ernst Mayr has shaped the lives of 20th-century scientists: through his ideas, his writings, his collaborations, his example, his lifelong warm friendships, and his encouragement.

Yet evolution needs to be understood not only by scientists, but also by the general public. Without understanding at least something of evolution, one has no chance of understanding the living world around us, human uniqueness, genetic diseases and their possible cures, and genetically engineered crops and their possible dangers. No other aspect of the living world is as fascinating and full of riddles as is evolution. How can one explain the remarkable adaptation of every species to its chosen niche? Or the beauty of birds of paradise, butterflies, and flowers? How can one explain the gradual advance from the simplest bacteria three-and-a-half billion years ago to dinosaurs, whales, orchids, and giant sequoias? Natural theologians had raised such questions for hundreds of years, but were unable to find any other answer than the hand of a wise and almighty creator. Eventually, Darwin argued that the fascinating world of life had gradually evolved by natural processes from the simplest kinds of bacteria-like organisms, and he backed up his claim by presenting a well-thought-out theory of evolution. Most importantly, he also proposed a theory of causation, the theory of natural selection.

Although the basic idea that evolution was responsible for biological diversity became widely accepted almost immediately after 1859, more specific aspects of evolution remained controversial for the next 80 years. Throughout this period there were constant disagreements about the causes of evolutionary change, about how species originated, and about whether evolution was a gradual or discontinuous process. The so-called Evolutionary Synthesis of 1937–1947 brought widespread consensus, and the molecular biology revolution in the ensuing years continued to strengthen the Darwinian paradigm and its support among biologists. Although numerous attempts were made in these years to propose opposing theories, not one of them has been successful: all have been thoroughly refuted.

Increasingly, it was realized that the Darwinian paradigm was important not only for explaining biological evolution, but more broadly for understanding our entire world and the human phenomenon. This led to a remarkable outburst of publications dealing with all aspects of evolution. By now, about a dozen convincing refutations of the claims of the creationists concentrate on presenting the massive evidence for the fact of evolution. Specialists can now consult three superb texts of evolutionary biology, those by Futuyma, Ridley, and Strickberger, which in more than 600 pages deal with all aspects of evolution in the utmost detail. These books provide an excellent education in the facts and theories of evolutionary biology.

Yet the available literature, excellent as it is, leaves a gap: our lack of a mid-level account of evolution, written not just for scientists but for the educated public, with special emphasis on explanations of evolutionary phenomenon and processes. This is the area where Ernst Mayr's *What Evolution Is* excels. We are lucky that, after a lifetime of writing for scientists, Ernst has now distilled his unparalleled experience for the public. Every major evolutionary phenomenon is treated as a problem that requires an explanation. Ernst often makes use of the history of failed earlier explanations to bring out the nature of the ultimate correct solution.

Also very helpful is Ernst's organization of the subject matter into three parts: (A) the evidence for evolution, (B) the explanation for evolutionary change and adaptation, and (C) the origin and meaning of biodiversity. A separate chapter, on the history of mankind, presents very successfully the evolution of humans and their precursors (hominids), which arose as "just another" group of apes. That chapter includes novel ideas, such as a suggested cause for the sudden drastic increase of brain size in the evolution from *Australopithecus* to *Homo*, and a suggested source of altruistic behavior.

For what audience is Ernst's *What Evolution Is* particularly suitable? One can answer: for the audience of everyone interested in evolution — particularly for anyone who really wants to understand the underlying causes of evolutionary change. Technical details, such as those dealing with the latest discoveries of molecular biology, are omitted because they can be found in detailed texts of evolution as well as in any modern biology text. *What Evolution Is* will be an ideal text in a course on evolution for non-biologists. Palaeontologists and anthropologists will welcome this book because of its emphasis on concepts and explanations. Ernst's lucid writing makes the subject of evolution accessible to any educated layperson.

Darwinism has become so fascinating in recent years that now every year at least one new book is published with the word "Darwin" in the title. It will greatly help the readers of such volumes to evaluate the claims made there by consulting *What Evolution Is*. Darwinian thinking, particularly the principle of "variation and selection (elimination)," is now widely employed in the humanities and social sciences. Those who employ it will find *What Evolution Is* a useful guide.

I can summarize my views on Mayr's *What Evolution Is* by saying that anybody with even the slightest interest in evolution should own and read this book. You will be richly rewarded. There is no better book on evolution. There will never be another book like it.

Jared M. Diamond

Evolution is the most important concept in biology. There is not a single Why? question in biology that can be answered adequately without a consideration of evolution. But the importance of this concept goes far beyond biology. The thinking of modern humans, whether we realize it or not, is profoundly affected—one is almost tempted to say determined—by evolutionary thinking. To offer a volume dealing with this important subject thus requires no apology.

However, someone might say, "Is not the market already saturated with books about evolution?" As far as the sheer quantity of published volumes is concerned, the answer might well be "Yes." Particularly there are several excellent technical texts for biologists who specialize in evolutionary studies. There are also splendid defenses of evolutionism against attacks by creationists, as well as excellent volumes on special aspects of evolution, such as behavioral evolution, evolutionary ecology, coevolution, sexual selection, and adaptation. But none of them quite fills the niche I have in mind.

This volume is meant for three kinds of readers. First and foremost, it is written for anyone, biologist or not, who simply wants to know more about evolution. Such a reader is quite aware how important this process is but does not understand exactly how it works and how one can answer some of the attacks against the Darwinian interpretation. The second group of readers consists of those who accept evolution, but are in doubt whether the Darwinian explanation is the correct one. I hope to answer all the questions this kind of reader is apt to ask. And finally, my account is directed to those creationists who want to know more about the current paradigm of evolutionary science, if for no other reason than to be able to better argue against it. I do not expect to convert this kind of reader, but I want to show him or her how powerful the evidence is that induces the evolutionary biologist to disagree with the account presented in Genesis.

The existing volumes intended to fill these needs have some of the following shortcomings. All of them are rather poorly organized and fail to present a concise, reader-friendly account. Most of them are not as didactic as they should be, because a difficult subject such as evolution should be presented as answers to a series of questions. Nearly all of them devote too much space to specialized aspects of evolution, such as the genetic basis of variation and the role of sex ratios. Virtually all of them are too technical and use too much jargon. About one-quarter of the content of all recent major evolutionary texts is devoted to genetics. I agree that the principles of genetics must be thoroughly explained, but there is no need for so much Mendelian arithmetic. Nor should space be wasted on arguing for or against obsolete claims, such as that the gene is the object of selection, or to a refutation of extreme recapitulationism (the idea that ontogeny recapitulates or repeats phylogeny). On the other hand, several of these texts do not give adequate space to an analysis of the different kinds of natural selection, particularly selection for reproductive success.

Most existing volumes on evolution have two other weaknesses. First, they fail to point out that almost all evolutionary phenomena can be assigned to one or the other of two major evolutionary processes: the acquisition and maintenance of adaptedness, and the origin and role of organic diversity. Although both take place simultaneously, they must be analyzed separately for a full understanding of their respective roles in evolution.

Second, most treatments of evolution are written in a reductionist manner in which all evolutionary phenomena are reduced to the level of the gene. An attempt is then made to explain the higher-level evolutionary process by "upward" reasoning. This approach invariably fails. Evolution deals with phenotypes of individuals, with populations, with species; it is not "a change in gene frequencies." The two most important units in evolution are the individual, the principal object of selection, and the population, the stage of diversifying evolution. These will be the major objects of my analysis.

It is remarkable how often a person who is trying to solve a particular evolutionary problem goes through the same sequence of unsuccessful attempts to find the solution, as has the whole field of evolutionary biology in its long history. Let us remember that our current understanding of evolution is the result of 250 years of intensive scientific study. Anyone trying to understand the solution of a given evolutionary problem may be greatly helped by considering the steps (many of them unsuccessful) by which the valid answer was finally found. It is for this didactic reason that I frequently present in considerable detail the history of the advance toward the solution of a challenging problem. Finally, I pay particular attention to human evolution and discuss to what extent our improved understanding of evolution has affected the viewpoints and values of modern humans.

What I have aimed for is an elementary volume that stresses principles and does not get lost in detail. I try to remove misunderstandings, but do not devote excessive space to ephemeral controversies, such as the role of punctuated equilibria or neutral evolution. Also, there is no longer any need to present an exhaustive list of the proofs for evolution. That evolution has taken place is so well established that such a detailed presentation of the evidence is no longer needed. In any case, it would not convince those who do not want to be persuaded.

Ernst Mayr
Harvard University

ACKNOWLEDGMENTS

Having been interested in evolution since before the 1920s, most of what I have learned I owe to masters of evolutionary thought whom I can no longer thank in person. I think of Theodosius Dobzhansky, R. A. Fisher, J. B. S. Haldane, David Lack, Michael Lerner, B. Rensch, G. Ledyard Stebbins, and Ervin Stresemann. The list should actually be far longer, but these are names that came to my mind at this moment. They certainly were a group of powerful thinkers who constructed modern Darwinism.

It gives me great pleasure to thank personally numerous evolutionists who helped me in the preparation of this volume by supplying information or critical comment: Francisco Ayala, Walter Bock, Frederick Burkhardt, T. Cavalier-Smith, Ned Colbert, F. DeWaal, Jared Diamond, Doug Futuyma, M.T. Ghiselin, G. Giribet, Verne Grant, Steve Gould, Dan Hartl, F. Jacob, T. Junker, Lynn Margulis, R. May, Axel Meyer, John A. Moore, E. Nevo, David Pilbeam, William Schopf, Bruce Wallace, and E. O. Wilson, R. W. Wrangham, Elwood Zimmermann.

The librarians of the Ernst Mayr Library at the Museum of Comparative Zoology were most helpful in tracking literature references and other help with this bibliography. Deborah Whitehead, Joohee Lee, and Chenowoth Moffatt prepared the manuscript and contributed in numerous other ways to its completion. Doug Rand saved the electronic program of illustrations from impending disaster. Finally, I am most grateful to Basic Books and its editorial staff, particularly Jo-Ann Miller, Christine Marra, and John C. Thomas in guiding the manuscript through the editorial process.

i
WHAT IS
EVOLUTION?

CHAPTER 1

....

IN WHAT KIND OF A WORLD DO WE LIVE?

Mankind apparently has always had an urge to explain and understand that which is unknown or puzzling. The folklore of even the most primitive human tribes indicates that they had given some thought to questions about the origin and history of the world. They had thought about such questions as: Who or what gave rise to the world? What will the future bring? How did we humans originate? Numerous answers to these questions were given in tribal myths. Most often the existence of the world was simply taken for granted, as was the belief that it had always been as it is now, but there were innumerable stories about the origin or creation of man.

Later on the founders of religions, as well as the philosophers, also tried to find answers to these questions. When one studies these answers, one can sort them into three classes: (1) a world of infinite duration, (2) a constant world of short duration, and (3) an evolving world.

(1) A world of infinite duration

The Greek philosopher Aristotle believed that the world had always been in existence. Some philosophers thought that this eternal world had never changed, that it was constant; others thought that it was going through different stages ("cycling") but would ultimately always return to an earlier stage. However, such a belief in an infinite age of the world was never very popular. There seems to have been an urge to account for a beginning.

(2) A constant world of short duration

This was, of course, the Christian view, as presented in the Bible. It was the prevailing view of the Western world in the Middle Ages and up to the middle of the nineteenth century. It was based on a belief in a supreme being, an all-powerful God, who had created the entire world as well as the human species, as described in the two stories of creation in the Bible (Genesis).

The belief that the world was created by an Almighty God is called creationism. Most of those who hold this belief also believe that God designed his creation so wisely that all animals and plants are perfectly adapted to each other and to their environment. Everything in the world today is still as it was when it was created. This was an entirely logical conclusion based on the known facts at the time the Bible was written. Some theologians, on the basis of the biblical genealogy, calculated that the world was quite recent, having been created in 4004 B.C., that is, about 6,000 years ago.

The beliefs of creationism are in conflict with the findings of science, and this has resulted in a controversy between creationists and evolutionists. This book is not the place to settle their arguments and we refer to the extensive literature on this subject listed in Box 1.1 and the bibliography. For the source of the creation stories in Genesis, see Moore (2001).

More or less similar creation stories are found in the folklore of peoples all over the world. They filled a gap in mankind's desire to

---

Box 1.1 Anticreationist Books

Berra, Tim M. 1990. *Evolution and the Myth of Creationism*. Stanford: Stanford University Press.

Eldredge, Niles. 2000. *The Triumph of Evolution and the Failure of Creationism*. New York: W. H. Freeman.

Futuyma, Douglas J. 1983. *Science on Trial: The Case for Evolution*. New York: Pantheon Books.

Kitcher, Philip. 1982. *Abusing Science: The Case Against Creationism*. Cambridge, Mass.: MIT Press.

Montagu, Ashley (ed.). 1983. *Science and Creationism*. New York: Oxford University Press.

Newell, Norman D. 1982. *Creation and Evolution: Myth or Reality?* New York: Columbia University Press.

Peacocke, A. R. 1979. *Creation and the World of Science*. Oxford: Clarendon Press.

Ruse, Michael. 1982. *Darwinism Defended*. Reading, Mass.: Addison-Wesley.

Young, Willard. 1985. *Fallacies of Creationism*. Calgary, Alberta, Canada: Detrelig Enterprises.

answer the profound questions about this world that we humans have asked ever since there has been human culture. We still treasure these stories as part of our cultural heritage, but we turn to science when we want to learn the real truth about the history of the world.

THE RISE OF EVOLUTIONISM
....

Beginning with the Scientific Revolution in the seventeenth century, more and more scientific observations were in conflict with the biblical story. Its credibility was gradually being weakened by a series of discoveries. The Copernican Revolution was the first development to demonstrate that not every statement in the Bible could be interpreted literally. The newly developing science was at first primarily concerned with astronomy, that is, with the sun, the stars, the planets, and other physical phenomena. It was inevitable that in due time the early practitioners of science would feel compelled to find explanations for many other phenomena in the world.

Discoveries in other sciences also raised new puzzling questions. The research of geologists in the seventeenth and eighteenth centuries revealed the immense age of the Earth, while the discovery of extinct fossil faunas undermined the belief in the constancy and permanence of the Creation. Even though more and more evidence contradicted the assumption of the constancy of the world and its short duration, even though more and more voices were heard among scientists and philosophers questioning the validity of the biblical story, and even though the naturalist Jean-Baptiste de Lamarck had proposed in 1809 a full-fledged evolutionary theory, a more or less biblical worldview prevailed up to 1859, not only among laypeople but also among natural scientists and philosophers. It provided a simple answer to all questions about the world: God had created it and he had designed his created world so wisely that every organism was perfectly adapted to its place in nature. *The rise of evolution theory*

During this transitional period of conflicting evidence, all sorts of compromises were attempted to cope with these contradictions. One such attempt was the so-called *scala naturae*, the Great Chain of Being (Fig. 1.1), in which all entities in this world were arranged in an as-

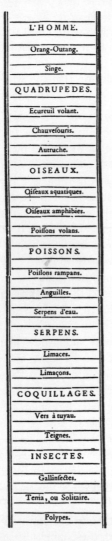

FIGURE 1.1

Great Chain of Being. Every variable found on Earth, from aspects of matter to animals and up to humans, was seen as a single continuous, linear "Great Chain" or *scala naturae*. Illustrated here is Bonnet's (1745) concept of this chain.

cending ladder, beginning with such inanimate objects as rocks and minerals, rising through lichens, mosses, and plants, and through corals and other lower animals to higher animals, and from them through mammals up to primates and man. This *scala naturae* was held to be never changing (constant) and simply to reflect the mind of the creator who had ordered everything in a sequence that led toward perfection (Lovejoy 1936).

Eventually, the evidence for the conclusion that the world is not constant but is forever changing became so overwhelming that it could no longer be denied. The result was the proposal and eventual adoption of a third worldview.

(3) An evolving world

According to this third view, the world is of long duration and is forever changing; it is evolving. Even though this may seem strange to us moderns, the concept of evolution was at first alien to Western thought. The power of the Christian fundamentalist dogma was so strong that it required a long series of developments in the seventeenth and eighteenth centuries before the idea of evolution became fully acceptable. As far as science is concerned, the acceptance of evolution meant that the world could no longer be considered merely as the seat of activity of physical laws but had to incorporate history and, more importantly, the observed changes in the living world in the course of time. Gradually the term "evolution" came to represent these changes. *Third view*

What Kinds of Change?

Everything on this Earth seems to be in a continuous flux. There are highly regular changes. The change from day to night and back again, caused by the rotation of the Earth, is such a regular cyclical change. So are the changes of the sea level in the tides, caused by the lunar cycle. Even more pervasive are the seasonal changes due to the annual circling of the Earth around the sun. Other changes are irregular, such as the movements of the tectonic plates, the severity of the winter from year to year, or aperiodic climatic changes (El Niño, ice ages), as well as periods of prosperity in a given nation's economy. Irregular changes are largely unpredictable, being subject to various stochastic processes.

There is, however, one particular kind of change that seems to keep going continuously and to have a directional component. This change is referred to as *evolution*. The first widespread feeling that the world was not static as implied by the story of Creation, but rather was evolving, can be traced to the eighteenth century. Eventually it was realized that the static *scala naturae* could be converted into a kind of biological escalator, leading from the lowest organisms to ever higher ones and finally to man. Just as gradual change in the development of an individual organism leads from the fertilized egg to the fully adult individual, so it was thought that the organic world as a whole moved from the simplest organisms to ever more complex ones, culminating in man. The first author to articulate this idea in detail was the French naturalist Lamarck. One even took the word evolution, which originally had been applied by Charles Bonnet to the development of the egg, and transferred it to the development of the world of life. Evolution, one said, consists of a change from the simple to the complex and from the lower to the higher. Evolution, indeed, was change, but it seemed to be a directional change, a change toward ever greater perfection, as it was said at that time, not a cyclical change like the seasons of the year or an irregular change like the ice ages or the weather.

But what is it that is actually involved in this continuing change of the organic world? This question was at first quite controversial, even though Darwin already knew the answer. Finally, during the evolutionary synthesis (see below), a consensus emerged: "Evolution is change in the properties of populations of organisms over time." In other words, the population is the so-called *unit of evolution*. Genes, individuals, and species also play a role, but it is the change in populations that characterizes organic evolution.

It is sometimes claimed that evolution, by producing order, is in conflict with the "law of entropy" of physics, according to which evolutionary change should produce an increase of disorder. Actually there is no conflict, because the law of entropy is valid only for closed systems, whereas the evolution of a species of organisms takes place in an open system in which organisms can reduce entropy at the expense of the environment and the sun supplies a continuing input of energy.

Evolutionary thinking spread throughout the second half of the eighteenth and the first half of the nineteenth century, not only in biology but in linguistics, philosophy, sociology, economics, and other branches of thought. Yet, on the whole, in science it remained for a long time a minority view. The actual shift from the belief in a static worldview to evolutionism was caused by the dramatic event of the publication of Charles Darwin's *On the Origin of Species* on the 24th of November in 1859.

DARWIN AND DARWINISM

This event represents perhaps the greatest intellectual revolution experienced by mankind. It challenged not only the belief in the constancy (and recency) of the world, but also the cause of the remarkable adaptation of organisms and, most shockingly, the uniqueness of man in the living world. But Darwin did far more than postulate evolution (and present overwhelming evidence for its occurrence); he also proposed an explanation for evolution that did not rely on any supernatural powers or forces. He explained evolution naturally, that is, by using phenomena and processes that everybody could daily observe in nature. In fact, in addition to the theory of evolution as such, Darwin proposed four theories about the how and why of evolution. No wonder the *Origin* caused such turmoil. It almost single-handedly effected the secularization of science.

Charles Darwin was born on February 12, 1809, the second son of a physician in a small English country town (Fig. 1.2). From his boyhood on, he was an ardent naturalist, particularly passionate about beetles. At his father's wish, he studied medicine in Edinburgh for a while, but was so appalled, particularly by the operations, that he soon gave it up. The family then decided he should study for the ministry, and this seemed a perfectly natural education for a young naturalist, for nearly all leading naturalists of his time were ordained ministers. Although Darwin conscientiously did all the required reading in the classics and in theology, it was really natural history that he pursued with single-minded devotion. After obtaining his degree at Cambridge University (Christ College), he received through one of his teachers at Cambridge the invitation to join one of the Navy's survey ships, HMS *Beagle*, for a survey of the coasts of South America,

FIGURE 1.2
Young Darwin at ca. 29 years, at the height of his intellectual creativity. *Source*: Negative no. 326694, courtesy the Library, American Museum of Natural History

particularly the harbors. The *Beagle* left England at the end of December 1831. On the five-year cruise of the *Beagle*, Darwin shared a cabin with the commander, Captain Robert Fitzroy. While the ship surveyed the coast of Patagonia in the east, the Strait of Magellan, and parts of the western coast and adjacent islands, Darwin had abundant opportunity to explore the mainland and the biota of the islands. Throughout the trip, he not only made significant collections of natural history specimens, but more importantly he asked endless questions about the history of the land and its fauna and flora. This was the foundation on which his evolutionary ideas grew.

After his return to England in October 1836, he devoted his time to the scientific study of his collections and to the publication of scientific reports, at first on some of his geological observations. After a few years, he married his cousin Emma, the daughter of the famous potter Wedgwood, bought a house near London (Down House), and lived there until his death on April 19, 1882, at the age of 73. It was at Down House that he wrote all of his major papers and books.

What made Darwin such a great scientist and intellectual innovator? He was a superb observer, endowed with an insatiable curiosity. He never took anything for granted but always asked why and how. Why is the fauna of islands so different from that of the nearest mainland? How do species originate? Why are the fossils of Patagonia basically so similar to Patagonia's living biota? Why does each island in an archipelago have its own endemic species and yet they are all much more similar to each other than to related species in more distant areas? It was this ability to observe interesting facts and to ask the appropriate questions about them that permitted him to make so many scientific discoveries and to develop so many highly original concepts.

Darwin also saw clearly that there are two aspects of evolution. One is the "upward" movement of a phyletic lineage, its gradual change from an ancestral to a derived condition. This is referred to as *anagenesis*. The other consists of the splitting of evolutionary lineages or, more broadly, of the origin of new branches (clades) of the phylogenetic tree. This process of the origin of biodiversity is called *cladogenesis*. It always begins with an event of speciation, but the new clade may become, in time, an important branch of the phylogenetic tree by diverging increasingly from the ancestral type. The study of cladogenesis is one of the major concerns of macroevolutionary research. Anagenesis and cladogenesis are largely independent processes (Mayr 1991).

Already in the 1860s knowledgeable biologists and geologists accepted that evolution was a fact, but Darwin's explanations of the how and why of evolution faced protracted opposition, as we shall show in later chapters. But let us first review some of the evidence for the actual occurrence of evolution that has been gathered since 1859.

. . . .

CHAPTER 2

WHAT IS THE EVIDENCE
FOR EVOLUTION ON EARTH?

The pre-Darwinian theories of evolution had little impact. Even though some evolutionary thinking was widespread among geologists, biologists, and even among literary people and philosophers, the Biblical story of Creation, as told in the book of Genesis, Chapters 1 and 2, was virtually unanimously accepted not only by laypeople but also by scientists and philosophers. This changed overnight, so to speak, in 1859 with the publication of Charles Darwin's *On the Origin of Species*. Even though some of Darwin's explanatory theories of evolution continued to encounter much resistance for another 80 years, his conclusion that the world had evolved was widely accepted within a few years after 1859.

However, throughout the nineteenth century whenever people talked about evolution, they referred to it as a theory. To be sure, at first, the thought that life on Earth could have evolved was merely a speculation. Yet, beginning with Darwin in 1859, more and more facts were discovered that were compatible only with the concept of evolution. Eventually it was widely appreciated that the occurrence of evolution was supported by such an overwhelming amount of evidence that it could no longer be called a theory. Indeed, since it was as well supported by facts as was heliocentricity, evolution also had to be considered a fact, like heliocentricity. This chapter will be devoted to a presentation of the evidence that led to the adoption of the "evolution is a fact" conviction among scientists. It will also challenge those who are still not yet convinced of the occurrence of evolution.

Evolution is a historical process that cannot be proven by the same arguments and methods by which purely physical or functional phenomena can be documented. Evolution as a whole, and the explanation of particular evolutionary events, must be inferred from observations. Such inferences subsequently must be tested again and again against new observations, and the original inference is either falsified or considerably strengthened when confirmed by all of these tests. However, most inferences made by evolutionists have by now been tested successfully so often that they are accepted as certainties.

WHAT EVIDENCE DOES THE EVOLUTIONIST HAVE?

The evidence for evolution is now quite overwhelming. It is presented in great detail by Futuyma (1983, 1998), Ridley (1996), and Strickberger (1996), and also in the anticreationist volumes listed in Chapter 1. My own treatment focuses on the classes of evidence now available to document evolution. It shows how remarkably congruent are the conclusions drawn from the most diversified branches of biology, which all support evolution. Indeed, these findings would make no sense in any other explanation.

The Fossil Record

The most convincing evidence for the occurrence of evolution is the discovery of extinct organisms in older geological strata. Some of the remnants of the biota that lived at a given geological period in the past are embedded as fossils in the strata laid down at that period. Each earlier stratum contains the ancestors of biota fossilized in the succeeding stratum. The fossils found in the most recent strata are often very similar to still living species or, in some cases, even indistinguishable. The older the strata are in which a fossil is found—that is, the further back in time—the more different the fossil will be from living representatives. Darwin reasoned that this is to be expected if the fauna and flora of the earlier strata had gradually evolved into their descendants in the later, more recent strata.

Given the fact of evolution, one would expect the fossils to document a gradual steady change from ancestral forms to the descendants. But this is not what the paleontologist finds. Instead, he or she finds gaps in just about every phyletic series. New types often appear quite suddenly, and their immediate ancestors are absent in the earlier geological strata. The discovery of unbroken series of species changing gradually into descending species is very rare. Indeed the fossil record is one of discontinuities, seemingly documenting jumps (*saltations*) from one type of organism to a different type. This raises a puzzling question: Why does the fossil record fail to reflect the gradual change one would expect from evolution?

All of his life Darwin insisted that this is simply due to the unimaginable incompleteness of the fossil record. Only an incredibly small fraction of organisms that had once lived are preserved as fossils. Often the fossil-bearing strata were on plates that were subsequently subducted and destroyed in the process of plate tectonics. Others were strongly folded, compressed, and metamorphosed, obliterating the fossils. Only a fraction of the fossil-bearing strata is presently exposed at the Earth's surface. But it is even highly improbable that any organism ever becomes fossilized at all, since most dead animals and plants are either eaten by scavengers or decay. They become fossilized only when, immediately after death, they are buried by sediment or volcanic ash. Fortunately, occasionally a rare fossil is found that fills the gap between ancestors and modern descendants. *Archaeopteryx*, for instance, a primitive fossil bird of the upper Jurassic (145 million years ago), still had teeth, a long tail, and other characteristics of his reptilian ancestors. However, in other respects, for instance in its brain, large eyes, feathers, and wings, it is rather similar to living birds. Fossils that fill a large gap are referred to as *missing links*. The discovery of *Archaeopteryx* in 1861 was particularly gratifying because anatomists had already concluded that birds must have descended from reptilian ancestors. *Archaeopteryx* confirmed their prediction.

A few fossil lineages are remarkably complete. This is true, for instance, for the lineage that leads from the therapsid reptiles to the mammals (Fig. 2.1). Some of these fossils appear to be so intermediate between reptiles and mammals that it is almost arbitrary whether to call them reptiles or mammals. A remarkably complete set of tran-

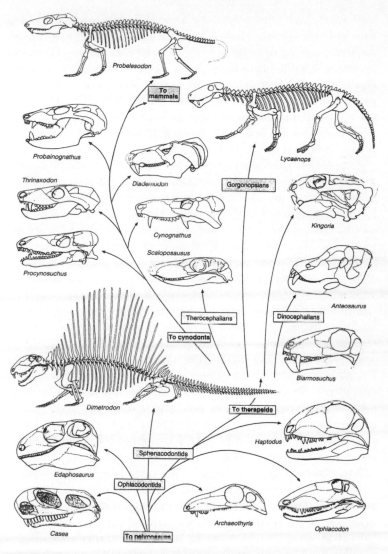

FIGURE 2.1

Evolution of the synapsid Reptilia, with the cynodonts forming a transition to the earliest mammals. *Source*: Ridley, M. (1993). *Evolution*. Blackwell Scientific: Boston, p. 535. Reprinted by permission of Blackwell Science, Inc.

sitions was also found between the land-living ancestors of the whales and their aquatic descendants. These fossils document that whales are derived from ungulates (mesonychid condylarthra) that increasingly became adapted to life in water (Fig. 2.2). The *australopithecine* ancestors of man also form a rather impressive transition from a chimpanzeelike anthropoid stage to that of modern man. The most complete transition between an early primitive type and its modern descendant that has been described is that between *Eohippus*, the ancestral horse, and *Equus*, the modern horse (Fig. 2.3).

The study of phylogeny is really a study of *homologous* characters. Since all members of a taxon must consist of the descendants of the nearest common ancestor, this common descent can be inferred only by the study of their homologous character. But how do we determine whether or not the characters of two species or higher taxa are homologous? We say that they are if they conform to the definition of homologous: *A feature in two or more taxa is homologous when it is derived from the same (or a corresponding) feature of their nearest common ancestor.*

This definition applies equally to structural, physiological, molecular, and behavioral characteristics of organisms. But how are we to determine whether homology is substantiated in a particular case? Fortunately, there are numerous criteria (see Mayr and Ashlock 1991). For structures this includes the position in relation to neighboring structures or organs; by connecting two dissimilar features by intermediate stages in ancestors; by similarity in ontogeny; and by intermediate fossils. The best evidence for homology has been provided in recent years by molecular biology. Such research has provided reliable evidence on the relationship of nearly all higher taxa of animals, and rapid progress is now also being made in reconstructing the relationship of the higher taxa of plants. A taxon, delimited by the methods of Darwinian classification, and therefore consisting exclusively of descendants of the nearest common ancestor, is called *monophyletic.*

What is particularly convincing about fossil animal series is that each fossil type is found at the time level at which one ought to expect it. For instance, modern mammals began to evolve after the Alvarez extinction event at the beginning of the Paleocene (60 million years ago). No modern mammal, therefore, should be found in strata that are 100 or 200 million years old, and indeed none has ever been

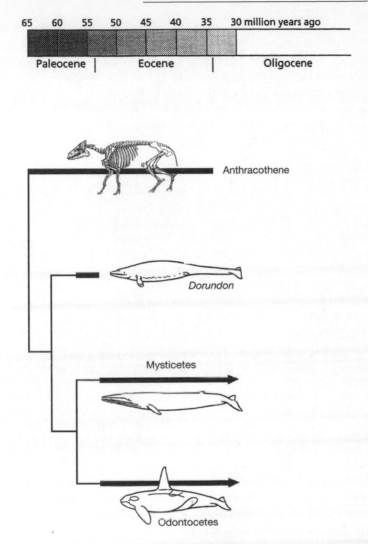

FIGURE 2.2

The descent of the whales from Eocene artiodactye ungulates is now reasonably well documented by transitional fossils. *Source*: From various sources, particularly personal information from Prof. Philip D. Gingerich

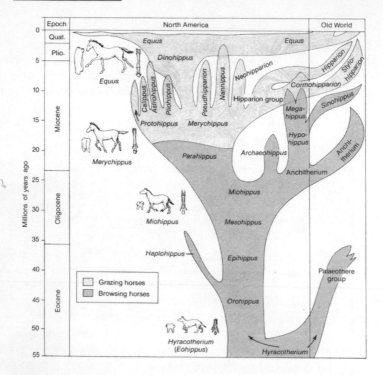

FIGURE 2.3

Evolution of the horse family, from the Eocene *Hyracotherium (Eohippus)* to the modern horse *(Equus)*. There was an extraordinary origin, flourishing, and extinction of
types of horses in the Miocene. *Source*: Strickberger, Monroe, W., *Evolution*, 1990,
Jones and Bartlett, Publishers, Sudbury, MA. www.jbpun. com. Reprinted with permission.

found. Or, to take another example, giraffes originated in mid-
Tertiary times about 30 million years ago. It would upset all our beliefs and calculations if suddenly a fossil giraffe was found from the
Paleocene 60 million years ago. But, of course, no such fossil has ever
been found.

Formerly, the ages of these fossils were mere guesses. All one knew
was that the lower strata were older than the higher strata. However,
the clock provided by the constancy of radioactive decay now permits
extremely precise age determinations of certain strata, particularly
lavas and other volcanic deposits that appear between fossil deposits

(see Box 2.1). Carbon dating can be used for the most recent past. The age of any fossil can now be determined with remarkable precision if one knows in what geological stratum it was found (Fig. 2.4). At the turn of the twenty-first century, the sequence of accurately dated fossils has documented evolution in the most convincing manner (see page 37).

BRANCHING EVOLUTION AND COMMON DESCENT
··

The *scala naturae* was a linear progression from lower to higher, and in Lamarck's presentation of evolution, each lineage originated with a (single-cell) *infusorian* believed to have originated by spontaneous generation. In the course of evolution its descendants became ever more complex and more perfect. Indeed, all pre-Darwinian evolutionary schemes postulated essentially straight phyletic lineages (see Chapter 4). One of Darwin's major contributions was to have proposed the first consistent theory of *branching evolution*.

It was an observation he made on the birds of the Galapagos Islands that led him to the branching theory. The Galapagos Islands are actually peaks of submarine volcanoes that have never had a land connection with South America or any other continent. All of the Galapagos fauna and flora got there by over-water (distance) colonization. Darwin knew that there was only one species of mockingbird in South America, but he found a species of mockingbird on each of three is-

---

Box 2.1   Radioactive Clock

Certain rocks, mostly of volcanic origin (e.g., lava flows), contain radioactive minerals such as potassium, uranium, and thorium. Each of these minerals decays at a specific rate and physicists have determined their half-lives. Uranium 238, for instance, has a half-life of 4.5 billion years, producing lead 206 in the process. The age of a given rock can then be calculated from the ratio of uranium and lead. Sedimentary rocks, which do not contain radioactive minerals, are dated by their location relative to datable strata.

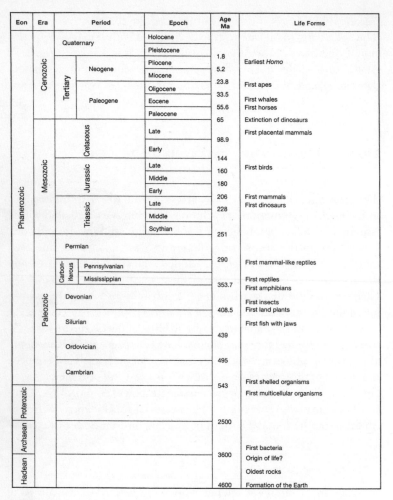

Eon	Era	Period		Epoch	Age Ma	Life Forms
Phanerozoic	Cenozoic	Quaternary		Holocene		
				Pleistocene		
		Tertiary	Neogene	Pliocene	1.8	Earliest *Homo*
				Miocene	5.2	
			Paleogene	Oligocene	23.8	First apes
				Eocene	33.5	First whales
				Paleocene	55.6	First horses
	Mesozoic	Cretaceous	Late		65	Extinction of dinosaurs
						First placental mammals
			Early		98.9	
		Jurassic	Late		144	First birds
			Middle		160	
			Early		180	
		Triassic	Late		206	First mammals
			Middle		228	First dinosaurs
			Scythian		251	
	Paleozoic	Permian			290	First mammal-like reptiles
		Carboniferous	Pennsylvanian			First reptiles
			Mississippian		353.7	First amphibians
		Devonian				First insects
					408.5	First land plants
		Silurian				First fish with jaws
					439	
		Ordovician			495	
		Cambrian			543	First shelled organisms
						First multicellular organisms
Proterozoic					2500	
Archaean					3600	First bacteria
						Origin of life?
Hadean						Oldest rocks
					4600	Formation of the Earth

FIGURE 2.4

The geological timescale. The Precambrian ranges from the origin of life (ca. 3,800 million years ago) to the beginning of the Cambrian (ca. 543 million years ago). New fossil finds frequently require a correction of the date of the earliest occurrence of a higher taxon. *Source: Evolutionary Analysis* 2nd ed. by Freeman/Herron, copyright © 1997. Reprinted by permission of Pearson Education, Inc., Upper Saddle River, NJ.

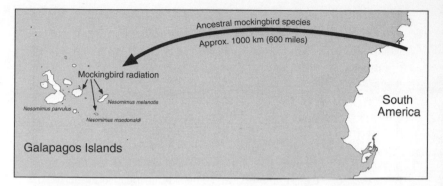

FIGURE 2.5

Colonization of the Galapagos Islands by an ancestral South American mockingbird species and its subsequent evolution into three local species.

lands in the Galapagos (Fig. 2.5), with each species different from the others. He concluded quite rightly that a single colonization of the South American mockingbird had given rise, by branching descent, to three different species on three different islands in the Galapagos. Then, he further reasoned that probably all mockingbirds in the world had descended from a common ancestor, because they are basically so similar to each other. Mockingbirds and their relatives, like thrashers and catbirds, then presumably also had a common ancestor.

This chain of inferences led Darwin to the ultimate conclusion that all organisms on Earth had common ancestors and that probably all life on Earth had started with a single origin of life. As Darwin wrote, "There is grandeur in this view of life, with its several powers, having been originally breathed into a few forms or into one; and that from so simple a beginning endless forms most beautiful and most wonderful have been, and are being, evolved" (1859: 490). As we shall presently show, numerous studies making use of different kinds of evidence have convincingly confirmed Darwin's conjecture. It is now referred to as the theory of *common descent.*

Paleontologists, geneticists, and philosophers had long been puzzled over how and where the branching took place that leads to the phenomenon of common descent. This problem was solved by taxonomists, who showed that it is speciation, particularly often geographic speciation, that leads to branching (see Chapter 9).

The theory of common descent solved a long-standing puzzle of natural history. There seemed to be a basic conflict between the overwhelming diversity of life and the observation that certain groups of organisms often shared the same characteristics. Thus there were frogs, snakes, birds, and mammals, yet the basic anatomy of all of these so different appearing classes of vertebrates was very much the same, yet totally different from that of an insect. The theory of common descent provided the answer to this puzzling observation. When certain organisms share a series of joint characteristics, in spite of numerous other differences, it is due to the fact that they had descended from the same common ancestor. Their similarities were due to the heritage they had received from this ancestor, and the differences had been acquired since the ancestral lines had split.

How Well Is Common Descent Documented?

The fossil record provides abundant evidence for common descent. For instance, in mid-Tertiary strata we may find fossils that are the common ancestors of dogs and bears. In somewhat earlier strata we find common ancestors of dogs and cats. Indeed, paleontologists have succeeded in showing that all carnivores descended from the same common ancestral type. The same descent from the common ancestor is true for all rodents, all ungulates, and for all other orders of mammals. Indeed, this principle of common descent also holds true for birds, reptiles, fishes, insects, and all other groups of organisms.

Even before 1859, zoologists had been able to construct a rather detailed taxonomic hierarchy of animal taxa. What was still not yet understood was why there was such a hierarchy. It was Darwin who showed that it could be explained by the principle of common descent. All the species of a genus have a nearest common ancestor and so do all the species of a family or of any other higher category in the hierarchy. This joint ancestry is the reason why the members of a taxon are so similar to each other.

Morphological Similarity.   Very suggestive evidence for common descent is also provided by the study of comparative anatomy. It was customary already in the eighteenth century to call certain organisms

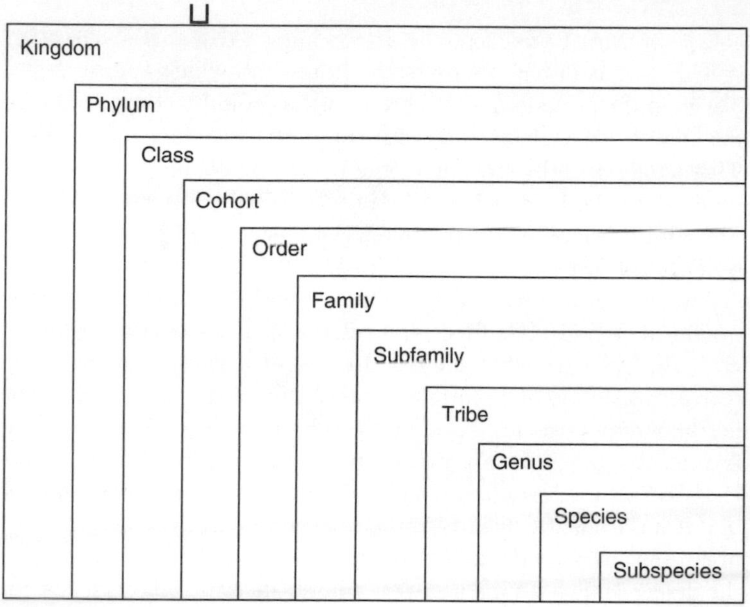

LJ

Kingdom
Phylum
Class
Cohort
Order
Family
Subfamily
Tribe
Genus
Species
Subspecies

FIGURE 2.6

The Linnacan hierarchy. Each category is nested within the next higher category, such as the species in the genus.

"related" when they were similar. At that time the French naturalist Comte Buffon described this for horses, donkeys, and zebras. The less similar that two kinds of organisms were, the less closely they were considered to be "related." The systematists, the students of classification, used the degree of similarity to establish a hierarchy of taxonomic categories. The most similar organisms were placed in the same species. Similar species were placed in the same genus, similar genera in the same family, and thus all the way up to the taxa of the highest category.

This arrangement of organisms by the degree of their similarity and relationship is called the *Linnaean hierarchy* (Fig. 2.6), after the Swedish botanist Carolus Linnaeus, who developed the system of binomial classification. Such a classification groups organisms into larger and larger taxa, finally comprising all the animals and all the plants. Beginning with a particular species, let us say the cat, one was able to construct this hierarchy. It was known that there were other species of cats rather similar to the house cat, which Linnaeus also placed in the genus *Felis*. This group of cats could be combined with the lion, the cheetah, and other genera of cats into the family Felidae. This family of catlike mammals could then be combined with other predatory mammals such as the Canidae (doglike), Ursidae (bears), Mustelidae (weasels), Viverridae (civets), and related groups into the order of Carnivora.

In a similar manner, other mammals could be combined into the orders of Artiodactyla (deer and relatives), Perissodactyla (horses, etc.), Rodentia (rodents, etc.), and those of whales, bats, primates, marsupials, and so on to form the class Mammalia (mammals). A similar hierarchy exists for all other kinds of animals, such as birds and insects, and for plants. The nature and causation of this grouping, unless ascribed to creation, was a complete riddle until Darwin showed that it was evidently due to "common descent." Each taxon (group of organisms), Darwin demonstrated, could be explained as consisting of the descendants from the nearest common ancestor, and such descent required evolution. The observed facts fit Darwin's theory of evolution so perfectly that his theory of "common descent by modification" was accepted almost immediately after 1859. Classification, a most active occupation of so many nineteenth-century zoologists and botanists, now had an explanation. The most frequently used evidence, on the basis of which relationship and common descent was inferred, was morphological and embryological similarity, and the search for such similarity led in the second half of the nineteenth century to a great flowering of comparative morphology and embryology (Bowler 1996).

Phylogeny, a special branch of biology, deals with the pattern and history of the descent of organisms. The pattern of descent is often presented as a phylogenetic tree (*dendrogram*) or in a certain school of taxonomists as a cladogram. Inspired by Ernst Haeckel, a German zoologist and contemporary of Darwin, zoologists and botanists have devoted much time and effort to clarify the actual phylogeny of organisms (see Chapter 3).

The Explanation of Morphological Types A second, related branch of biology likewise found its explanation through common descent. The comparative anatomists, led by Georges Cuvier, had recognized a limited number of types of organisms that agreed with each other in their basic structure (archetype). Cuvier (1812) distinguished four major phyla *(embranchements)*, all members of which, he thought, had the same *Bauplan* (body plan). The existence of these very distinct types, not connected by any intermediates or transitions, decisively refuted the validity of the *scala naturae*. Cuvier called these types Vertebrates, Mollusks, Articulates, and Radiates. This was a first step, but it was soon shown that three of his types were composite, while the vertebrates were ultimately classed as a subdivision of the Chordates. At the present time, about 30 phyla of animals are recognized, and in most of them several minor types are distinguished, for example, in the vertebrates there are fishes, amphibians, reptiles, birds, and mammals. Again, the existence of these morphological types made sense as soon as one recognized that each consisted of the descendants of a common ancestor who shared its basic body plan.

The preevolutionary morphologists, like Cuvier, were typologists (essentialists) in their thinking. They were followers of Plato. Each type (phylum) was considered to be completely separated from the others, it was defined by its essence, and it was constant. Even though the philosophical basis of this so-called idealistic morphology was quite wrong, its emphasis on the study of morphology led to numerous discoveries of great value for the reconstruction of phylogeny and, more broadly, for the understanding of evolution.

Homology It is quite remarkable how successful comparative morphology can be in the reconstruction of missing steps in an evolutionary sequence. T. H. Huxley, for instance, when reconstructing the nonflying ancestor of birds, concluded that it was an archosaurian reptile. *Archaeopteryx*, a remarkable bridge between birds and the archosaurians, was discovered only a few years later, in 1861. Evolutionary entomologists postulated that ants had evolved from wasplike ancestors and inferred what characters the earliest ants must have had. When a fossil ant was then discovered in mid-Cretaceous amber, it largely confirmed the inferred reconstruction. These are not isolated cases, for whenever a missing ancestor was reconstructed, it agreed remarkably well with the real ancestor subsequently discovered as fossil.

During evolution any characteristic of an organism may be modified. Yet, even in preevolutionary days, some comparative anatomists realized which modified structures were equivalent, such as the wings

Homologous
explains evolution and how
we have a common ancestor

of birds and the anterior extremities of mammals. Richard Owen, a typological morphologist, said such structures were "homologous" and defined them as "the same organ in different animals under every variety of form and function." This, of course, left it wide open how to decide when two organs were "the same organ." This problem was solved by Darwin, who said that certain characteristics of two species were homologous if they were derived by evolution from an equivalent characteristic in the nearest common ancestor of the two species. The anterior extremity of a walking mammal, let us say a dog, was appropriately modified by evolution for such different functions as digging (mole), climbing (monkey), swimming (whale), and flying (bat) (Fig. 2.7). Furthermore, this mammalian structure is homologous with the pectoral fin of certain fishes.

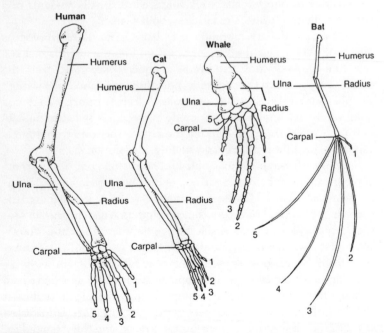

FIGURE 2.7
Adaptive modifications of mammalian forelimbs. The homologous bone elements of human, cat, whale, and bat have been modified by selection to serve their species-specific functions. *Source*: Strickberger, Monroe, W., *Evolution*, 1990, Jones and Bartlett, Publishers, Sudbury, MA. www.jbpub. com. Reprinted with permission.

The claim that certain characteristics in rather distantly related taxa are homologous is at first merely a conjecture. The validity of such an inference must be tested by a series of criteria (Mayr and Ashlock 1991), such as position in relation to neighboring organs, the presence of intermediate stages in related taxa, similarity of ontogeny, existence of intermediate conditions in fossil ancestors, and agreement with evidence provided by other homologies. Homology cannot be proven; it is always inferred.

Homology is due to the partial inheritance of the same genotype from the common ancestor. This is the reason why homology exists not only for structural characters, but for any inheritable feature, such as behavior. Characters that have independently arisen by parallelophyly are nevertheless homologous, because they were produced by the genotype of the common ancestor. Homologous structures may differ considerably in their development. For a review of the different ways in which the term homology has been used, see Butler and Saidel (2000).

Embryology.   Perceptive anatomists observed in the eighteenth century that the embryos of related kinds of animals are often far more similar to each other than are the adult forms. An early human embryo, for instance, is very similar not only to embryos of other mammals (dog, cow, mouse), but in its early stages even to those of reptiles, amphibians, and fishes (Fig. 2.8). The older the embryo, the more it shows the special characters of the higher taxon to which it belongs. When the adults are highly specialized (for instance, the sessile barnacles among the crustaceans) their free-swimming larvae are still very similar to those of other crustaceans (Fig. 2.9). Some of Darwin's opponents asserted that such larval similarities would prove nothing. All development by necessity moves from simple to complex, they said, and the early developmental stages, being simpler, are thus more similar than the later, more complex ones. This is in part true, but embryos and larvae always have some characteristics peculiar to the phyletic lineage to which they belong, and thus reveal their relationship. Furthermore, study of the embryonic stages very often shows how a common ancestral stage gradually diverges in different branches of the ancestral tree. This leads to a far better understanding of the evolutionary pathways.

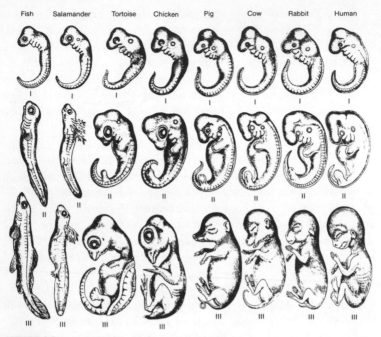

Fish Salamander Tortoise Chicken Pig Cow Rabbit Human

FIGURE 2.8

Haeckel's figure of 1870 showing the similarity of the development of human embryos to three comparable stages in 7 other kinds of vertebrates. Haekel had fraudulently substituted dog embryos for the human ones, but they were so similar to humans that these (if available) would have made the same point. Source: Strickberger, Monroe, W., *Evolution*, 1990, Jones and Bartlett, Publishers, Sudbury, MA. www.jbpub.com. Reprinted with permission.

Recapitulation \ The term "recapitulation" refers to the appearance and subsequent loss of structures in ontogeny, which in related taxa are retained in the adults. Thus it refers to the loss of an ancestral character in later embryonic stages in one phyletic lineage, but the retention of this character in living species of other lineages derived from the same common ancestor. For instance, embryos of the baleen whales still develop teeth at certain embryonic stages, but these are later reabsorbed and disappear. This appearance and subsequent loss of ancestral characters in succeeding embryonic stages is so striking a phenomenon that it led to a special theory, that of *recapitulation*. Two drastically different interpretations of these observations were offered by the embryologists.

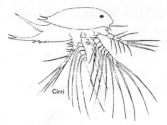

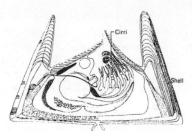

(c) Nauplius. *(Redrawn from Costlow in Etkin and Gilbert.)*

(c) *Balanus*, a crustacean. *(Redrawn from Barnes.)*

FIGURE 2.9

The free-swimming larval stages of barnacles (cirripedia) are like those of other crustaceans, but the sessile adult stages are so different that the early zoologists considered them to be molluscs. *Source*: Kelly, Mahlon G. and McGrath, John C. (1975). *Biology: Evolution and Adaptation to the Environment*. Houghton Mifflin.

According to Karl Ernst von Baer's theory, embryos of different organisms in their earliest embryonic stages are so similar to each other that they can not be correctly identified unless one knows their origin. But during development they gradually become more and more similar to their adult stage and, thus, diverge increasingly from the developmental lineage of other organisms. Von Baer summarized this view in his well-known statement: "There is gradually taking place a transition from something homogeneous and general to something heterogeneous and special." His explanation was widely accepted. However, this claim was clearly in conflict with certain facts of ontogeny. For instance, why should the embryos of birds and mammals develop gill slits, like fish embryos? Gill slits are not a more general condition of the neck region of a terrestrial vertebrate (see Fig. 2.8). These embryonic gill slits had been discovered in the 1790s, that is, 70 years before the publication of the *Origin*. At that time, the only available explanation was the Great Chain of Being, the *scala naturae*, in which all organisms were arranged in a series of ever greater "perfection" from the lowest organism to fish, reptiles, and eventually man. This led to the suggestion that the embryo of a higher organism "recapitulated" the ontogeny of organisms lower on the *scala naturae*. When evolution was accepted, a new definition offered by Haeckel (1866), "Ontogeny is the recapitulation of phylogeny," obviously went too far, because at no stage of its development does a mammalian embryo look like an adult

Homogeny   vs   ontogeny.

fish. Yet, in certain features, as in the gill pouches, the mammalian embryo does indeed recapitulate the ancestral condition. And such cases of recapitulation are by no means rare. The larvae of barnacles are very similar to those of other crustaceans (Fig. 2.9), and embryonic structures are found in thousands of cases to be indicative of their ancestry, but these same structures are absent in the adult life-forms.

The embryologist could not escape the question of why in these cases ontogeny followed such a roundabout way to reach the adult stage, instead of simply eliminating the embryonic structures that are no longer needed, just as many cave-dwelling species eliminate pigmentation and eyes. The reason was eventually discovered by experimental embryologists, who found that these ancestral structures serve as embryonic "organizers" in the ensuing steps of development. For instance, if one cuts the pronephric duct of an amphibian embryo, there will be no development of the mesonephros. In a similar manner, the removal of the midline stripe of the archenteron roof prevents the development of a notochord and a nervous system. Thus the "useless" pronephros and midline stripe are recapitulated because they have the vital function of being embryonic organizers of later developing structures. This is the same reason why all terrestrial vertebrates (tetrapods) develop gill arches at a certain stage in their ontogeny. These gill-like structures are never used for breathing, but instead are drastically restructured during the later ontogeny and give rise to many structures in the neck region of reptiles, birds, and mammals. The evident explanation is that the genetic developmental program has no way of eliminating the ancestral stages of development and is forced to modify them during the subsequent steps of development in order to make them suitable for the new life-form of the organism. The anlage of the ancestral organ now serves as a somatic program for the ensuing development of the restructured organ (Mayr 1994). What is recapitulated are always particular structures, but never the whole adult form of the ancestor.

Vestigial Structures.   Many organisms have structures that are not fully functional or not functional at all. The human caecal appendix is an example, and so are the teeth in baleen whale embryos and the eyes in many cave animals. Such vestigial structures are the remnants of structures that had been fully functional in their ancestors but are now greatly reduced owing to a change in niche utilization. When these structures lose their function owing to a shift in lifestyle, they are no longer protected by natural selection and are gradually deconstructed. They are informative by showing the previous course of evolution.

These three phenomena—embryonic similarities, recapitulation, and vestigial structures—raise insurmountable difficulties for a creationist explanation, but are fully compatible with an evolutionary explanation based on common descent, variation, and selection.

Biogeography. Evolution also helped to explain another great puzzle of biology, namely, the reasons for the geographic distribution of animals and plants. Why are the faunas of Europe and North America on both sides of the North Atlantic so relatively similar, whereas those of Africa and South America on both sides of the South Atlantic are so very different? Why is the fauna of Australia so strikingly different from that of all other continents? Why are there normally no mammals on oceanic islands? Could these seemingly capricious patterns of distribution be explained as the product of creation? Not easily. Darwin, however, showed that the present distribution of animals and plants is due to the history of their dispersal from their original points of origin. The longer that two continents were isolated from each other, the more different their biota became.

Many organisms have what is called discontinuous distributions. For example, camels and their relatives are found on two different continents: the true camels in Asia and Africa, and their close relatives the llamas in South America. If we believe in continuous evolution there should be a connection between the two now isolated areas; in other words, camels should occur in North America, but they are absent. This situation led to the inference that camels had indeed at one time existed in North America, serving as a connecting link between the Asian and South American camels, but then had become extinct. In due time, this conjecture was indeed confirmed by the discovery in North America of a large fossil fauna of Tertiary camels (Fig. 2.10). Likewise, the reasons for the similarity of the fauna of Europe and of North America were not fully understood until it was discovered that in the early Tertiary (40 million years ago) there was a broad land connection across the North Atlantic between the two now-separated

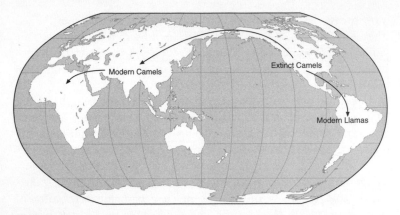

FIGURE 2.10

The ranges (Asia to South America) of the existing members of the camel family are widely separated. The discovery of a rich fossil fauna of camels in the Tertiary of North America showed that once there was a complete faunal continuity.

continents. This permitted an active faunal exchange. By contrast, Africa and South America were separated by continental drift 80 million years ago and their biota diverged greatly during their long isolation. Again and again, puzzling distribution patterns can be explained as the result of common descent and sometimes subsequent extinction. Thus evolution continues to provide the answer to many previously puzzling observations.

Dispersal Different species can have highly divergent dispersal abilities. More than 100 species of New Guinea birds are so averse to crossing water gaps that they are not found on a single island more than one mile distant from the mainland coast. On the other hand, some species have truly miraculous dispersal facilities. The lizard family Iguanidae is confined to the Americas, except for one genus (with two species) found in Fiji and Tonga (Fig. 2.11). Since these are endemic species they could not have been brought there by humans. The only possible explanation is that a long time ago they floated there on logs and flotsam carried by ocean currents. It is indeed almost unbelievable that these colonists should have been able to survive such a trip of several thousand miles. Even if at first they had only reached eastern Polynesia, where they were since exterminated by the Polynesians, it still was an extraordinary achievement. How-

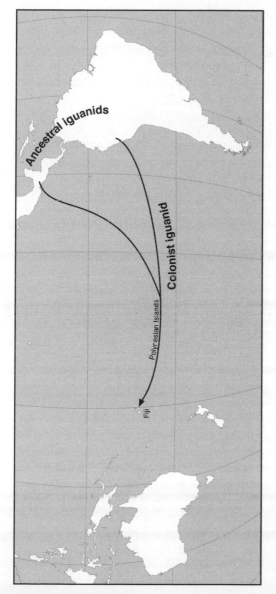

FIGURE 2.11

Extreme achievement of dispersal. The reptilian family Iguanidae is found only in North and South America, except for the two species of the endemic genus *Brachylophus* found thousands of kilometers away, in western Polynesia (Fiji, Tonga). It could have reached the Polynesian islands from the Americas only by rafting.

ever, there is no other explanation, and other cases of long survival on rafts are documented.

Differences in dispersal ability explain most of the apparent problems of distribution. Mammals (except bats) are notoriously ineffective in crossing water gaps, which is why they are usually absent from oceanic islands. This is also the reason why Wallace's Line in the Malay archipelago, a line between the Greater Sunda Islands in the west and the Lesser Sunda Islands and Sulawesi in the east, is an important biogeographic border for mammals, but much less so for birds and plants (Fig. 2.12). Actually this line separates the edge of the Sunda shelf from deep water to the east. Mammals are restricted to the land of the Sunda shelf, while many birds and plant seeds can cross water gaps with considerable ease.

Distributional Gaps   The ranges of some taxa are broken by a gap in which the taxon does not occur. There are two different ways by which such gaps may originate. The North American gap in the range of the camel families, as we saw above, was caused by their extinction there. Originally they ranged continuously from Asia to South America. This is referred to as the vicariance hypothesis. Most discontinuities on continents seem to be such remnants of previously continuous ranges. Many arctic species, for instance, were able at the height of the Pleistocene glaciation to colonize the Alps and Rocky Mountains, but are now left after the retreat of the ice as montane relics, widely separated from the arctic populations of their species.

A second type of range discontinuity is primary. It originates when members of a species establish a founder population beyond the present species border after dispersing across unsuitable terrain (water, mountains, or an unsuitable vegetation area). Such dispersal discontinuities are particularly characteristic for areas with insular distributions. The taxa of the Galapagos Islands never had a continuous range with South America, their source area. All species of this insular biota reached the Galapagos Islands by crossing the 600-mile water gap between the two areas. For a creationist there is no rational explanation for distributional irregularities, but they are completely compatible with a historical evolutionary explanation.

Molecular Evidence.   It was one of the unexpected happy discoveries of molecular biology that molecules evolve just the same as do so-

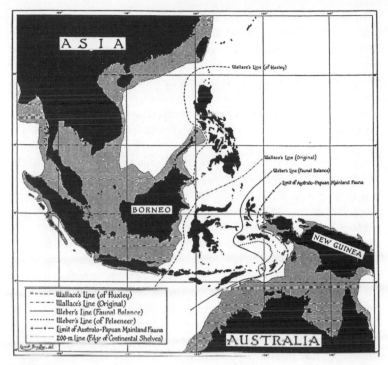

FIGURE 2.12

The contact zone between the Indo-Malayan and the Australo-Papuan faunas. The shaded area in the west is the Asian (Sunda) shelf, and in the east lies the Australian (Sahul) shelf. The area between the two shelves, never connected by a land bridge, is referred to as Wallacea. The real border (line of balance) between the Asian and the Australian faunas is Weber's Line. *Source*: Mayr, Ernst. (1944). *Quarterly Review of Biology* 19(1): 1–14.

matic structures. On the whole, the more closely related two organisms are, the more similar are their respective molecules. In many cases when there was considerable doubt as to the relationship of two organisms because the morphological evidence was ambiguous, a study of their molecules revealed the real relationship. As a result, molecular biology became one of the most important sources of information on phylogenetic relationships.

Genes, or more precisely the structure of the molecules of which they consist, undergo an evolutionary change just as do macroscopic structures. By comparing homologous genes and other homologous molecules of different organisms, one can determine the degree of their similarity. However, different kinds of molecules have different rates of evolutionary change. Some change very rapidly, like the fibrinopeptides, and others change very slowly, like the histones. Even though the lineages of man and chimpanzee separated at least 6 million years ago, the highly complex molecules of the hemoglobins of these two species are still virtually identical. What is gratifying is the fact that when a phylogeny based on morphological or behavioral characteristics is established, it is usually found to be essentially the same as a phylogeny based exclusively on molecular characteristics.

A comparison of the results of both sources of evidence is most helpful in all cases in which the analysis of morphology has led to ambiguous results. Such cases can now be tested against the molecular phylogeny of these taxa. Many different genes are available for such an analysis. In some cases it is the molecular evidence that reflects the phylogeny more correctly than the morphology. To mention just two cases from the recent literature, molecular analysis showed that the golden mole of South Africa and the tenrecs of Madagascar are quite unrelated to the Insectivora, among which these animals were traditionally classified on the basis of morphological evidence. Likewise, the Pogonophora and the Echiura, always considered independent phyla, were shown to be more closely related to certain families of polychaetes than these are to other polychaetes. The extremely close relationship of man to the chimpanzee and to the other anthropoid apes is as convincingly documented by molecular as by structural characters.

The Importance of Molecular Analysis One of the most important contributions made by molecular biology to the understanding of evolution is the discovery that the basic molecular framework of all organisms is very old. The particular structures acquired by the phyla of animals, fungi, and plants that enable them to survive and prosper in the particular niche or adaptive zone that they occupy are, on the whole, considerably more recent. So we can use these adaptive structures to classify animals, fungi, and plants, but they tell us little about how the fungi are related to animals or plants. For instance, fungi traditionally were always considered to be related to plants and their study was the job of botany departments. To be sure, it was puzzling that their cell walls consisted of chitin, a substance supplying all the hard parts of insects but not found anywhere in plants. This was simply treated as one of the typical exceptions that are so common in biology. But molecular analysis finally showed that in much of their basic chemistry fungi are quite closely related to the Animalia.

The gradual straightening out of the chaos of the 50–80 phyla of "protists" is also a great achievement of molecular biology (and of the study of membranes and other fine structures), after a study of the traditional morphological characters had failed to produce clarity. The successful arrangement of the angiosperms into groups of related orders and families was likewise largely accomplished by the application of molecular methods. Perhaps the greatest virtue of the molecular approach is that there are so many potential characters to study. When one particular gene leads to ambiguous results, one can in principle shift to any of thousands of other genes to test a suspected connection.

The Molecular Clock In the absence of an adequate fossil record, for a long time it was essentially impossible to determine the geological age of many evolutionary lineages. However, Zuckerkandl and Pauling (1962) showed that many, perhaps most, molecules have a rather constant rate of change over time. Such molecules can serve as a *molecular clock*. Well-dated fossils with modern descendants provide us with a yardstick for calibrating a given molecular clock. It was by the molecular clock method that the branching point between chimpanzee and man was shown to be as recent as 5–8 million years ago, rather than 14–16 million years, as had been previously generally accepted.

However, the molecular clock method must be applied with caution because molecular clocks are not nearly as constant as often believed. Not only do different molecules have different rates of change, but a particular molecule may vary its rate over time. These represent cases of *mosaic evolution*. In cases of discrepancy it is always advisable to determine also the rate of change of a different molecule and to try to find another suitable fossil. Not always reliable

The Evolution of the Genotype as a Whole With the help of greatly improved methods it is now possible to determine the essentially complete DNA sequence of the entire genome of a whole organism. This was first done for several bacteria (eubacteria and archaebacteria), including *Escherichia coli*, then for yeast *(Saccharomyces)*, a plant

TABLE 2.1 Genome Size and DNA Content

Organism	Genome Size (base pairs x 10^9)	Coding DNA
Bacterium *(Escherichia coli)*	0.004	100
Yeast *(Saccharomyces)*	0.009	70
Nematode *(Caenorhabditis)*	0.09	25
Fruit fly *(Drosophila)*	0.18	33
Newt (Triturus)	19.0	1.5–4.5
Human (Homo sapies)	3.5	9–27
Lungfish (Protopterus)	140.0	0.4–1.2
Flowering plant (Arabidopsis)	0.2	31
Flowering plant (Fritillaria)	130.0	0.02

SOURCE: From Maynard Smith and Szathmary (1995), p. 5.

(Arabidopsis), and some animals, such as the roundworm (nematode) *Caenorhabditis* and the fruit fly *Drosophila* (Table 2.1). The completion of the essential sequencing of the human genome was celebrated in June 2000. The field dealing with the molecular structure of the genome is called *genomics*.

These sequences are now the material for the most fascinating comparative studies. Although genes (base pair sequences) evolve, the function of a gene sets severe limits on the amount of change. In other words, the basic structure of a gene is usually preserved over many millions of years and this permits the study of the phylogeny of each gene. The most astonishing result of these studies is that some basic genes of higher organisms can be traced all the way back to homologous genes in bacteria. Many genes in the yeast *Saccharomyces*, the worm *Caenorhabditis*, and the fly *Drosophila* can be traced back to the same ancestral gene. Such a gene may not have exactly the same function in all the organisms in which it occurs, but it will have a similar or equivalent function.

The Origin of New Genes   Bacteria and even the oldest eukaryotes (protists) have a rather small genome (see Box 3.1). This raises the question: By what process is a new gene produced? This occurs, most frequently, by the doubling of an existing gene and its insertion in the chromosome in tandem next to the parental gene. In due time the new gene may adopt a new function and the ancestral gene with its traditional function will then be referred to as the *orthologous* gene. It is through orthologous genes that the phylogeny of genes is traced. The derived gene, coexisting with the ancestral gene, is called *paralogous*. Evolutionary diversification is, to a large extent, effected by the production of paralogous genes. The doubling sometimes affects not merely a single gene, but a whole chromosome set or even an entire genome.

CONCLUSIONS

As we have seen, whatever aspect of biology is studied, it provides irrefutable evidence in support of evolution. As the famous geneticist T. Dobzhansky has said so rightly, "Nothing in biology makes sense, except in the light of evolution." Indeed, there is no other natural explanation than evolution for the facts presented in this chapter.

Perhaps nowhere has the evolutionary approach produced more clarity and understanding than in the ordering of the bewildering diversity of living organisms. As a result we can now describe in remarkable detail the gradual rise of higher organisms (plants and animals) from the simplest forms of life. The next chapter is devoted to a presentation of this ascent of life.

Astronomical and geophysical evidence indicate that the Earth originated about 4.6 billion years ago. At first the young Earth was not suitable for life, owing to heat and exposure to radiation. Astronomers estimate that it became livable about 3.8 billion years ago, and life apparently originated at about that time, but we do not know what this first life looked like. Undoubtedly, it consisted of aggregates of macromolecules able to derive substance and energy from surrounding inanimate molecules and from the sun's energy. Life may well have originated repeatedly at this early stage, but we know nothing about this. If there have been several origins of life, the other forms have since become extinct. Life as it now exists on Earth, including the simplest bacteria, was obviously derived from a single origin. This is indicated by the genetic code, which is the same for all organisms, including the simplest ones, as well as by many aspects of cells, including the microbial cells. The earliest fossil life was found in strata about 3.5 billion years old. These earliest fossils are bacteri-alike, indeed they are remarkably similar to some blue-green bacteria and other bacteria that are still living (Fig. 3.1).

THE ORIGIN OF LIFE
. . . .

What else can we say about the first beginnings of life? After 1859 some of Darwin's critics said: "This Darwin may well have explained

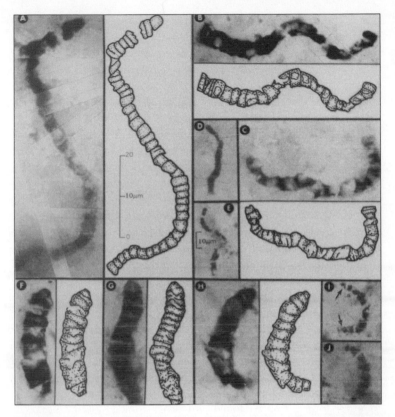

FIGURE 3.1

Fossil bacteria. Apparently the oldest are as much as 3,500 million years old and did not change much until modern times. *Source*: Reprinted with permission from J. Williams Schopf, "Microfossils of the early Archean Apex chert: New evidence of the antiquity of life, *Science* 260: 620–646, 1993. Copyright 1993 by the American Association for the Advancement of Science.

the evolution of organisms on earth, but he has not yet explained how life itself may have originated. How can inanimate matter suddenly become life?" This was a formidable challenge to the Darwinians. Indeed, for the next 60 years, this seemed an unanswerable question even though Darwin himself had already perceptively speculated on this issue: "all the conditions for the first production of a living organism . . . [could be met] . . . in some warm little pond with all sorts of ammonia and phosphoric salts, light, heat, electricity, etc. present" (Darwin 1859). Well, it did not turn out to be as easy as Darwin thought.

The Biosphere

From the origin of life on, there has been a dynamic interaction between organisms and their inanimate environment, particularly the atmosphere. The atmosphere of the young Earth was a reducing one (oxygen-free) consisting largely of methane, molecular hydrogen, ammonia, and water vapor. Eventually it was converted into an oxygen-containing atmosphere through the activity of blue-green bacteria (cyanobacteria). Limestones and other rock formations are further evidence of the effect of organisms (e.g., coral reefs) on the environment. There is often a steady-state balance of interaction between the activities of organisms and the responses of the inanimate environment. Interaction among different kinds of organisms likewise has a profound effect on the biosphere. Increased CO_2 production by a flourishing animal population will permit an increased CO_2 uptake by the plant world. The oxygen-rich atmosphere was apparently instrumental in the origin and success of the complex descendants of the prokaryotes—the eukaryotes. This interaction sometimes results in such a balanced steady state that some authors have proposed a *Gaia hypothesis*, according to which the Earth's inanimate and living worlds together form a well-balanced and programmed system. There is, however, no well-substantiated evidence for the existence of such a "program" and most evolutionists reject the Gaia hypothesis. They attribute the seeming balance to an opportunistic response of the living world to changes in the inanimate world and vice versa.

The first serious theories on the origin of life were proposed in the 1920s (Oparin, Haldane). In the last 75 years, an extensive literature dealing with this problem has developed and some six or seven competing theories for the origin of life have been proposed. Although no fully satisfactory theory has yet emerged, the problem no longer seems as formidable as at the beginning of the twentieth century. One is justified to claim that there are now a number of feasible scenarios of how life could have originated from inanimate matter. To under-

stand these various theories requires a good deal of technical knowledge of biochemistry. To avoid burdening this volume with such detail, I refer the reader to the special literature dealing with the origin
of life (Schopf 1999; Brack 1999; Oparin 1938; Zubbay 2000).

The first pioneers of life on Earth had to solve two major (and
some minor) problems: (1) how to acquire energy and (2) how to
replicate. The Earth's atmosphere at that time was essentially devoid
of oxygen. But there was abundant energy from the sun and in the
ocean from sulfides. Thus growth and acquisition of energy were apparently no major problem. It has even been suggested that rocky
surfaces were coated with metabolizing films that could grow but not
replicate. The invention of replication was more difficult. DNA is
now (except in some viruses) known as the molecule that is indispensable in replication. But how could it ever have been coopted for this
function? There is no good theory for this. However, RNA has enzy-
matic capacities and could have been selected for this property, with
its role in replication being secondary. It is now believed that there
may have been an RNA world before the DNA world. There was apparently already protein synthesis in this RNA world, but it lacked
the efficiency of the DNA protein synthesis.

In spite of all the theoretical advances that have been made toward
solving the problem of the origin of life, the cold fact remains that no
one has so far succeeded in creating life in a laboratory. This would
require not only an anoxic atmosphere, but presumably also other
somewhat unusual conditions (temperature, chemistry of the
medium) that no one has yet been able to replicate. It had to be a liquid (aqueous) medium that was perhaps similar to the hot water of the
volcanic vents at the ocean floor. Many more years of experimenta-
tion will likely pass before a laboratory succeeds in actually producing
life. However, the production of life cannot be too difficult, because it
happened on Earth apparently as soon as conditions had become suitable for life, around 3.8 billion years ago. Unfortunately we have no
fossils from the 300 million years between 3.8 and 3.5 billion years
ago. The earliest known fossiliferous rocks are 3.5 billion years old
and already contain a remarkably rich biota of bacteria. We have no
idea (and in the absence of fossils quite likely never will have) what
their ancestors in the preceding 300 million years looked like.

THE RISE OF ORGANIC DIVERSITY
....

Prokaryotes

Life on Earth originated ca. 3,800 million years ago. The earliest organisms were prokaryotes (bacteria), first encountered as fossils in strata that are 3,500 million years old. For the next 1,000 million years life on Earth consisted of prokaryotes. They differ from higher organisms, the eukaryotes (organisms with nucleated cells), by a large set of characters, best represented as the absence of the diagnostic characters of the eukaryotes (see Box 3.1). The bacteria are exceedingly diverse, with names such as cyanobacteria, gram-negative and gram-positive bacteria, purple bacteria, and archaebacteria. How they are related to each other and how they are to be classified is still rather controversial.

There are two major reasons for this failure of agreement. First, bacteria have neither biological species nor sexual reproduction. Instead they exchange genes and sometimes whole blocks of genes by a process called *lateral transfer*. A bacterium, for instance, may on the whole belong in a particular subdivision of the bacteria, let us say the gram-negative bacteria, but have a particular set of genes from an entirely different subdivision of the bacteria. It is therefore difficult, and may in some cases be entirely impossible, to construct the neat hierarchical trees found in the eukaryotes. The second reason for controversy is that the disagreeing specialists adhere to two very different taxonomic philosophies. The traditional classification of the prokaryotes followed the traditional principle of arranging all taxa on the basis of their degree of difference. Others instead follow the Hennigian ordering system in which taxa are arranged according to the sequence of branching points in the phylogenetic tree.

This dispute affects particularly the ranking of the archaebacteria. The Archaebacteria, a group of bacteria discovered by Woese, differ quite strikingly from the other bacteria in a few characters, especially in the cell wall and structure of the ribosomes. However, in all other characters they are typical prokaryotes. Indeed, a leading specialist in bacterial classification, Cavalier-Smith (1998), ranks the archaebacte-

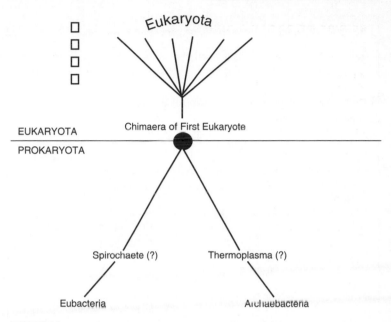

FIGURE 3.2
Model of the origin of the first eukaryote, through the formation of a chimaera between two prokaryotes, a eubacterium and an archaebacterium.

ria as one of the four subdivisions of the bacteria. They are no more different from the other three kinds of bacteria, he says, than are most subdivisions of the protists from each other. To be sure, they share with the eukaryotes the structure of the ribosomes and a few other characters. However, the first eukaryote originated by a symbiosis of an archaebacterium and a eubacterium and then by a chimaera formation of the two symbionts (Fig. 3.2). This is why the new taxon Eukaryotes combines characteristics of both archaebacteria and eubacteria (see Box 3.1).

It is difficult to determine which bacteria participated in this process. Spirochaetes must have been involved to provide the cilia. Lynn Margulis believes that five different bacterial genomes are recognizable in a simple protist. No doubt the first chimaera acquired additional genomes by unilateral gene transfer. The frequency of this

Box 3.1   Differences Between Prokaryotes and Eukaryotes

The number of known differences between prokaryotes and eukaryotes is now about 30. The few differences between archaebacteria and other bacteria pale by comparison.

Property	Prokaryotes	Eukaryotes
Cell size	Small, ca.1-10 pm	Large, normally 10-100 pm
Nucleus	Absent, nucleoid	Present (membrane bounded)
Endoplasmic membrane system	Absent	Endoplasmic reticulum, Golgi apparatus present
DNA	Not complexed with proteins	Organized into chromosomes with > 50% histones and/or other proteins
Organelles	No membrane-bound organelles	Normally contain organelles (mitochondria, chloroplasts, etc.)
Metabolism	Diverse	Aerobic (except for amitochondriats).
Cell wall	Peptidoglycan (protein) in eubacteria	Cellulose or chitin; none in animals
Reproduction	By binary fission, budding	Sexual via meiosis-fertilization cycles in animals and plants
Cell division	By fission	By mitosis
Genetic recombination	By unilateral gene transfer	By recombination during meiosis
Flagella	Rotating, made of flagellin proteins	Undulating cilia, made largely of tubulin
Respiration	On membranes	Mitochondria
Environmental tolerance	Euryoec	Stenoec
Propagules	Spores (endo- and exo-) resistant to desiccation; heat-resistant endospores	Great variety among phyla: cysts, seeds, etc.; less resistant to heat and desiccation than bacteria
Spliceosomes, peroxisomes, hydrogenosomes	Absent	Present

transfer, including that between such distantly related prokaryotes as eubacteria and archaebacteria, will make it very difficult to reconstruct the phylogeny of the prokaryotes.

The origin of the eukaryotes was arguably the most important event in the whole history of life on Earth. It made the origin of all the more complex organisms, plants, fungi, and animals possible. Nucleated cells, sexual reproduction, meiosis, and all the other unique properties of the more advanced multicellular organisms are achievements of the descendants of the first eukaryotes.

Prokaryotes remained exceedingly abundant after the origin of the eukaryotes and may have become even more abundant owing to their lifestyle of living on organic detritus and as parasites. According to some calculations the total biomass of the prokaryotes on Earth is as great as that of all the eukaryotes.

Bacteria have a large number of shared properties by which they differ from the eukaryotes, the "higher" organisms (Box 3.1): no nucleus; DNA located in gonophores; no protein-coated chromosomes; no sexual reproduction; cell division by simple fission or budding, but no mitosis or meiosis; bacterial flagella composed of flagellin protein, and flagella rotate; cells usually small (1–10 µm), some in colonial aggregates; and no cellular organelles (mitochondria, etc.).

Specialists differ on how the rich world of prokaryotes should be subdivided. One subdivision, the Archaebacteria, includes genera adapted to extreme environmental conditions, such as hot springs, sulfur springs, and brine, but others are found elsewhere, including ocean water.

The earliest fossil prokaryotes (3.5 billion years ago) were cyanobacteria (see Fig. 3.1). What is most remarkable about the cyanobacteria is their morphological stasis. About a third of the early fossil species of prokaryotes are morphologically indistinguishable from still living species and nearly all of them can be placed in modern genera. There are a number of possible reasons for this constancy. They reproduce asexually, they have very large populations, and they are able to live under highly variable and often extreme environmental conditions. All this may favor stability.

Eukaryotes

After about 1,000 million years of exclusively bacterial life on Earth, perhaps the most important and dramatic event in the history of life took place—the origin of the eukaryotes. Eukaryotes differ strikingly from prokaryotes by the possession of a nucleus surrounded by a membrane and containing individual chromosomes. The origin of the first eukaryote was a major evolutionary step. What apparently happened was the formation of a chimaera through symbiosis between an archaebacterium and a eubacterium to produce the first eukaryote (see Fig. 3.2). This mode of origin is inferred from the partly archaebacterial, partly eubacterial composition of the eukaryotic genome (Margulis et al. 2000). The new eukaryotic cell subsequently acquired various symbionts as cellular organelles, such as mitochondria and (in plants) chloroplasts. These organelles were probably acquired sequentially because some primitive living eukaryotes lack mitochondria or other intracellular organelles. It is not yet understood how the nucleus originated, in which the chromosomes are placed within a membrane. Symbiosis was apparently not involved in its origin.

The mitochondria were derived from the alpha subdivision of the purple bacteria (proteobacteria) and the chloroplasts of plants from cyanobacteria. The sequence of the processes by which the first eukaryotes were put together and their nucleus was acquired is still controversial. A spectacular new theory of the formation of the nucleus (Martin and Müller 1998) requires further testing before it can be considered a probable explanation.

Protists. The fossil record of the earliest eukaryotes is extremely poor. However, lipids (steranes), by-products of eukaryotic metabolism, have been recently discovered in rocks that are 2,700 million years (my) old. Thus the origin of the eukaryotes apparently occurred much earlier than formerly believed. There is, however, a very small possibility that these molecules had percolated down to these old sedimentary strata from more recent strata, but most geologists deny this possibility. The amount of free oxygen also increased at about that time and this appears to have greatly stimulated the rise of the eukaryotes. Molecular clock studies also support an early date for the origin of the eukaryotes. The early eukaryotes consisted of a single nucleated cell, with or without cellular organelles, and even though the unicellular eukaryotes are a very heterogeneous lot, they are usually collectively referred to in the vernacular as protists. Yet they are classified into a number of different kingdoms (Protozoa, Cnemista, etc.), and the simplest representatives of all higher taxa—plants, fungi, and animals—are also unicellular. Some of the protists that now lack intracellular organelles seem to have lost them secondarily.

After their origin ca. 2,700 my ago, the eukaryotes diversified spectacularly. The diversity of the protists is indicated by the fact that Margulis and Schwartz (1998) recognize no less than 36 phyla of protists. This includes amoebas, microsporidia, slime molds, dinoflagellates, ciliates, sporozoa, cryptomonads, flagellates, xanthophyta, diatoms, brown algae (some highly multicellular), oomycota, myxospora (sporozoa), red algae, green algae, radiolaria, and about 20 less well known phyla. Yet our incomplete understanding of the relationship among the unicellular eukaryotes is indicated by another modern classification that divides the protists into 80 phyla. The formal taxon Protista is no longer recognized owing to the extreme heterogeneity of the protists. It is evident that we are still a long way from a stable classification of the protists, which will require a far more intensive application of molecular methods.

The earliest fossils of unicellular eukaryotes (protists and algae) date to about 1,700 my, but various methods permit us to infer that they actually originated ca. 1,000 my earlier. The diversity of the early eukaryotes seemingly remained rather low for the period from 1,700 to 900 million years ago, but then rose rapidly to experience a veritable explosion of protistan microfossils during the Cambrian.

Multicellularity.   Multicellularity originated repeatedly during evolution. There are many forerunners of multicellularity among the bacteria. It seems that the first step toward multicellularity is an increase of size such as found in more than a dozen groups of unicellular protists, algae, and fungi. This usually leads to a division of labor among the cells of such aggregations, eventually merging into genuine multicellularity.

The earliest eukaryotes consisted of a single cell. Indeed, for a long time the protists were defined as unicellular eukaryotes. However, it was found that there are unicellular plants (green algae), unicellular animals (protozoans), and unicellular fungi. Furthermore, taxa that largely consisted of unicellular species, like the brown algae (Phaeophyta) and the red algae (Rhodophyta), also contain some highly multicellular species. The giant kelp (*Macrocystis*), which reaches a length of up to 100 m, belongs to a protist family. Some forms of multicellularity are widespread among basically unicellular taxa. Even bacteria sometimes aggregate into large masses of cells. Multicellularity reached its culmination in the three great kingdoms of Plants (metaphyta), Fungi, and Animals (metazoans). Older classifications recognized taxa of unicellular plants (algae), fungi, and animals (protozoans), but these unicellular organisms have now all been assigned to the protists.

THE PHYLOGENY OF THE ANIMALIA

The reconstruction of the animal phylogeny has long been controversial. The linear *scala naturae* of the eighteenth century was broken up by Cuvier in preevolutionary days into four phyla: vertebrates, molluscs, articulates, and radiates (Chapter 2). It was soon realized that Cuvier's radiates, composed of the coelenterates and echinoderms, was an artificial assemblage, and his other phyla were eventually reassorted step by step. The multicellular animals were finally classified as about 30 to 35 distinct "phyla." These phyla are the major groupings of animals, such as sponges, coelenterates, echinoderms, arthropods, annelids, molluscs, flatworms, and chordates, as well as numerous smaller phyla. All of them are separated from each other by a more or less pronounced gap. After 1859 it became the task of the evolutionist to determine how these phyla are related to each other and how they can be arranged in a single phylogenetic tree. What were the first multicellular animals like and which higher taxa gave rise to other still higher taxa? Students of phylogeny have been actively engaged in this search since the 1860s, and although the evolution of the animals is now understood in its major outlines, many details are still controversial. The seemingly most helpful arrangements are based on the traditional principles of Darwinian classification. Taxa are delimited on the basis of similarity, rather than at branching points.

Almost all of these phyla appeared seemingly full-fledged in the late Precambrian and early Cambrian, ca. 565–530 million years ago. No fossils intermediate between them have been found and no living intermediates are in existence. As a result, these phyla seem to be separated by unbridgeable gaps. How can these gaps be explained and how could they be bridged? A tentative explanation will be presented below. Since the earliest animals did not leave a fossil record, their phylogeny must be reconstructed through a study of their living descendants. A careful comparison of the morphology and embryology of the invertebrates led, after 100 years, to a reasonably robust construction of a phylogenetic tree of the animals. However, the relationship of several of the minor phyla remains uncertain and there is not yet even a complete consensus on some basic issues. Convergence, parallel evolution, extreme specialization, mosaic evolution, the loss of important characters, and other evolutionary phenomena seemed for a while to stymie any further advance. This impasse was broken when molecular characters were added to the morphological evidence.

When it was discovered that the molecules that make up genes undergo evolution and have a phylogeny just like morphological characters, it was hoped that a definite phylogeny of organisms could soon be constructed; molecular evidence would enable a decision whenever the morphological data were ambiguous. Alas, things did not turn out to be quite so simple, for this reasoning ignored the phenomenon of mosaic evolution. Each component of the genotype can evolve somewhat independently of the rest of the genotype. Endeavors to construct phylogenetic trees on the basis of the evolution of one particular molecule frequently produced results that were clearly in conflict with a massive amount of morphological and other evidence. For technical reasons the molecules that were first used for such analyses were ribosomal RNA and mitochondrial DNA. Unfortunately, these molecules often went their own evolutionary way. Tree phylogenies based on 18S RNA proved to be particularly misleading. In all more recent molecular analyses the conclusions are based on the study of several molecules, including nuclear genes. The occasional failures do not diminish the extraordinary contribution made by the molecular evidence. Building on the foundation of the solid achievements made by morphology and embryology, this new evidence now permits us to construct a well-tested phylogeny of the animal kingdom (Fig. 3.3).

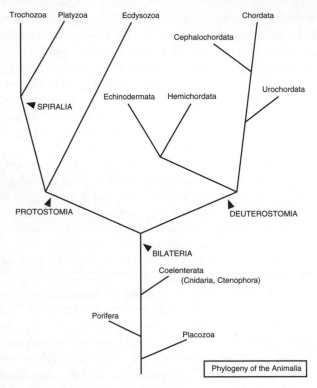

FIGURE 3.3
Proposed phylogeny of the major groups of animals. See text for the grouping of the phyla of Protostomia. Some of the tentative groupings are still controversial.

One can confidently predict that virtual consensus on animal phylogeny will be achieved within the next 15 years. Even now very few phyla are left whose position is still considered completely uncertain.

From the Earliest Animal to the Bilateria

The most primitive living multicellular animal is *Trichoplax (Placozoa)*, consisting largely of a ventral and a dorsal cell layer. It reproduces by "swarmers." Next higher are the sponges (Porifera), whose protistan ancestors seem to be the choanomonads. Molecular analysis suggests that the coelenterates, the next step in animal evolution, were derived from sponges. However, it is also possible that the coelenterates had originated independently from some group of protists. The two phyla of coelenterates (Cnidaria and Ctenophora) have a radially symmetrical morphology. Their embryos have two cell layers, an ectoderm and an endoderm; they are *diploblastic*. All other multicellular animals (Bilateria) are bilaterally symmetrical and have a third cell layer, the mesoderm; they are *triploblastic*.

The Evolution of the Bilateria

The relationships among the phyla of Bilateria have been controversial for more than 100 years. What classification was chosen before the introduction of molecular analysis depended entirely on the weight one gave to various morphological characters. The presence or absence of a coelom was long considered—erroneously—to be the most important character. The flatworms (Platyhelminthes) without a coelom were then considered the basic group of the Bilateria, giving rise to various derived groups. This is still a widely adopted (and well-supported) arrangement, but an alternative view that the platyhelminthes are a derived group who secondarily lost both coelom and anus is now also widely supported.

The Coelom. The earliest Bilateria are entirely soft-bodied. They crawl on the floor of the ocean or of other bodies of water. The other taxa of Bilateria derived from them can tunnel through the substrate not only for protection but also to exploit the rich sources of nutriment available in this niche. Peristaltic contractions of a strong mesodermal muscle sheet permit them to push through the soft substrate. This mode of propulsion is made possible by the squeezing pressure of the muscles of the body wall on cavities in the body filled with liquids. In some phyla, blood between the body tissues serves as the needed liquid. In most others, there are particular liquid-filled cavities, the so-called *coelom*. This hydrostatic system, consisting of the muscles of the body wall and the coelom, provides the needed rigidity for peristaltic locomotion.

Protostomia and Deuterostomia   The next step in the rise of the animals is the split of the Bilateria into two lineages, the Protostomia and the Deuterostomia. The blastopore in the gastrula stage of the developing embryo of a protostomian develops into the mouth opening of the adult, and the anus forms anew at the end of the gastrula sac, whereas in a deuterostome the permanent mouth is a newly formed opening and the blastopore becomes the anus (see Box 3.2). Furthermore, these two branches of animals differ in the coelom formation. The split between protostomes and deuterostomes is a very basic division of the animals.

The annelids, molluscs, arthropods, and a number of smaller phyla form the protostomes, while the echinoderms and chordates (including the vertebrates), together with three smaller phyla, form the deuterostomes. These two major groups differ by a number of fundamental characteristics. The development of the fertilized egg proceeds in most protostomia by spiral cleavage, in which the plane of cell division is diagonal to the vertical axis of the embryo. The deuterostome egg develops by radial cleavage (Fig. 3.4). However, some protostomia (e.g., Ecdysozoa) also develop by radial cleavage. The cleavage of most protostome eggs is *determinate*, that is, the ultimate function (role) of each part of the zygote is determined from the

Box 3.2	Differences Between Protostomia and Deuterostomia	
Property	*Protostomia*	*Deuterostomia*
Blastopore	Becomes adult mouth	Newly formed
Anus	Is newly formed	Formed from the blastopore
Coelom	If present, formed by schizocoely	Formed by enterocoely
Cleavage of fertilized egg	Usually spiral	Always radial
Development	Determinate	Indeterminate
Larvae	When present, with downstream collecting ciliary bands	Larvae with upstream ciliary bands

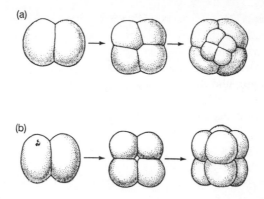

FIGURE 3.4

Spiral(a) vs. radial(e) pattern of the first cleavage division of the fertilized egg. *Source*: *Evolutionary Analysis*, 2nd ed. by Freeman/Herron, copyright © 1997. Reprinted by permission of Pearson Education, Inc., Upper Saddle River, NJ.

beginning. By contrast, in most deuterostomes the cleavage is *indeterminate*, that is, cells produced by early cleavage divisions retain the capacity to develop into a complete embryo.

As long as one had to depend on morphological characters, it remained controversial which phyla one should assign to the protostomians and which to the deuterostomes. Even more uncertain was the question of how to subdivide the protostomians with its numerous phyla. Molecular analysis has brought considerable clarity to these problems. A number of mathematical methods have now been developed to permit a translation of the molecular information into branching points between phyletic lineages. The methodology devoted to discover the branching pattern of phylogeny is called *cladistic* (or genealogical) *analysis*. Only derived characters provide information useful for the discovery of branching points.

About 24 phyla are usually recognized among the Protostomia. The uncertainty lies in whether some of the smaller taxa such as the Pogonophora, Echiura, and Micrognathozoa deserve phylum rank or should rather be ranked as classes or subphyla. The placement of most of the phyla is reasonably widely accepted, but for some phyla, such as the Chaetognatha, it is still rather uncertain. The following list of the phyla of Protostomia is widely accepted but cannot be considered final.

Ecdysozoa
  Panarthropoda
    Onychophora
    Tardigrada
    Arthropoda
  Introverta
    Kinorhyncha
    Priapulida
    Loricifera
    Nematoda
    Nematomorpha
Spiralia
  Platyzoa
    Gastrotricha
    Plathelminthes (or Platyhelminthes)
    Gnathostomulida
    Micrognathozoa
    Rotifera–Acanthocephala
    Cycliophora
    Chaetognatha
  Trochozoa (=Lophotrochozoa)
    Brachiopoda
    Bryozoa
    Phoronida
    Entoprocta
    Sipuncula
    Mollusca
    Annelida (including Pogonophora)
    Echiura
    Nemertea

Tentatively the Protostomia can be divided into two major groups of phyla, the Ecdysozoa and the Spiralia. All members of the Ecdysozoa undergo molt (ecdysis). This includes all arthropods and the nematodes and their relatives, thus some of the most species-rich phyla of animals. Two major groups of Spiralia are recognized. Those that have a lophophore feeding apparatus (bryozoans, brachiopods) and those that develop via trochophore larvae (annelids, molluscs, and others). Rotifers and relatives, nemertines, and platyhelminthes are tentatively placed here.

Most new phyla originate by "budding," that is, they originate as a side branch of one of the major phyla and often become so different in a relatively short time that their relationship is only discovered by molecular analysis. The derivation of some of the phyla is still somewhat uncertain.

The application of molecular methods has led to one important discovery: Complex characters, such as segmentation, the coelom, spiral cleavage, and a trochophore larva, are not such decisive proofs of relationship as had always been assumed, because they can again be lost in the course of evolution. For instance, much evidence indicates that the ancestors of the molluscs and Pogonophora were segmented and that those of the Platyhelminthes had a coelom. The presence of certain characters in the Pogonophora had long suggested their relationship to the Polychaetes even though this was not supported by other characters, now inferred to have been lost by the Pogonophora. Fortunately, the molecular characters give unambiguous answers in most of the cases of apparently lost characters.

The analysis of the characters of each of these phyla has consistently revealed that they all descended from common ancestors. For instance, the arthropods and annelids descended from an ancestral protostomian. The protostomes and deuterostomes derived from an ancestral bilateralian. Animals, plants, and fungi derived from ancestral single-celled eukaryotes, the eukaryotes from ancestral bacteria, and these from a single origin of life.

One may consider this taxonomic detail as rather uninteresting. For the evolutionist, however, it illuminates the steps by which the now existing organic diversity evolved. Certain branching events in the past have led to groups as distinct from each other as the Protostomia and the Deuterostomia, and retained the diagnostic difference between these taxa, while in other cases the same character (e.g., segmentation of the body) was acquired and lost several times in the course of time. A survey of the present diversity of higher taxa, and the success in tracing this diversity back to a limited number of ancestors, provides an impressive picture of the pathway of evolution.

Chronology of Animal Evolution. Not very many years ago the oldest known fossil animals were from the latest Precambrian about 550 my ago. The rich radiation of the animals was then thought to have taken place in the incredibly short time of only 10–20 million years. This seemed unbelievable and indeed has now been shown to have been wrong.

At first, all life on Earth lived in water. The first land plants date from about 450 million years ago, and the first flowering plants (angiosperms) from the Triassic, more than 200 million years ago. Insects, now the most species-rich group of higher organisms, originated at least 380 million years ago. Although the chordates originated ca. 600 million years ago, land vertebrates (amphibians) are first found in strata 460 million years old. Soon they gave rise to reptiles and the latter, more than 200 million years ago, to birds and mammals.

THE COMING AND GOING OF PHYLA
....

Geologists recognize definite periods (eras) in the history of the Earth. Each of these eras is characterized by the flowering or extinction of particular groups of organisms. The Cambrian (beginning 543 million years ago) is the age of the first major flowering of multicellular eukaryotes. The entire preceding history of the Earth is referred to as the Precambrian (4.6 billion to 543 million years ago). For at least 1 billion years after the inferred date of the origin of life (3.8 billion years ago), only prokaryotes existed. However, some time in the Proterozoic period (2.7 to 1.7 billion years ago) the eukaryotes originated and soon afterward the first multicellular eukaryotes. Even though they have not left a fossil record, their early date of origin can be inferred from the advanced evolution of their Cambrian descendants and by evolutionary clock calculations. The Ediacaran fauna of the latest Precambrian (650–543 million years ago) is the first fossil animal fauna.

The fossil-rich time span from the Cambrian to the present is called the Phanerozoic eon. Paleontologists divide it into the Paleozoic, Mesozoic, and Cenozoic eras. Each of these three eras is again subdivided into smaller periods. The break between the Paleozoic and Mesozoic is marked by the occurrence of a mass extinction at the end of the Permian period, and that between the Mesozoic and Cenozoic by a mass extinction at the end of the Cretaceous period.

The Origin of Multicellular Animals, the Cambrian Explosion

For a long time it was thought that the origin of multicellular animals had occurred in the Cambrian, which began 543 million years ago. In a short period the majority of the skeleton-bearing phyla of animals appeared as fossils in early Cambrian strata. Brachiopods, molluscs, arthropods (trilobites), and echinoderms were among the types that appeared at that time. The seeming suddenness of the simultaneous appearance of so many phyla of animals is perhaps only an artifact of another evolutionary development at that time. Most of the new fossils were discovered because they had a skeleton that their soft-bodied ancestors did not have. But then an even earlier fossil fauna (Ediacara) was discovered in various parts of the world in the late Precambrian (Vendian), containing many strange types as well as others clearly related to the Cambrian types. Some of the animals of this earlier Vendian fauna cannot be assigned to any of the now existing animal phyla, but these all became extinct before the Cambrian. The earliest triploblastic fossils of this fauna are dated 555 million years old.

If, as it seems probable, the apparent explosion of new phyla in the early Cambrian was in part due to the skeletonization of a great variety of already existing soft-bodied types, one would have to ask, what caused this sudden skeletonization of so many unrelated phyla? Two answers are usually given. There may have been a change in the Earth's atmosphere (e.g., increase of oxygen level) and in the chemistry of the seawater, or there may have been an evolution of efficient predators (requiring protection by an exoskeleton), or both.

This period of seemingly exuberant production of new structural types (phyla) soon came to an end. Altogether some 70 or 80 different structural types (body plans) appeared in the late Precambrian and early Cambrian, but apparently no new ones originated at any later period. To be sure, some small, soft-bodied taxa were first found as fossils in later periods, but their absence from the Cambrian is evidently only a matter of lack of preservation. Six phyla of now living small invertebrates have never been found as fossils.

All currently living phyla of animals, about 35 of them, were long thought to have originated during a period of only about 10 million years in the early Cambrian. How could one explain such short-lived exuberance of structural innovation at that period? Recent research indicates that this question is at least in part an artifact of the fossil record. A reconstruction of the date of origin of animal phyla with the help of the molecular clock methodology reveals a far earlier origin than indicated by the fossil record. Even though it is known that the molecular clock can occasionally speed up considerably, the molecular evidence requires that we adopt a much earlier date of origin of animal phyla than the Vendian (Precambrian) age. On the basis of the differences of 18 protein-coding gene loci, Ayala et al. (1998) estimate that the protostomes diverged from the deuterostomes about 670 my ago and the chordates from the echinoderms about 600 my ago. Coelenterates and sponges originated even earlier, one might guess at least 800 my ago.

Throughout this Precambrian period the rich diversity of protists gave rise to multicellular descendants, some of which then led to plants, fungi, and animals. In spite of much extinction, the dominant groups, which now characterize life on Earth, evolved at that time. Great age is also indicated by the complexity of some of the Cambrian fossils, which would have required evolution over hundreds of millions of years. The absence of the ancestral types in Precambrian strata can be explained if one assumes that the earliest multicellular animals were microscopically small and soft-bodied. Not only would they not be fossilized, but owing to their small size they would not even have left tracks on or in the substrate.

Yet, in addition to this factor, the early evolution of the metazoans may nevertheless have been exceptionally rapid. The genotype of the earliest metazoans may not have been as tightly constrained by regulatory genes as it is in their later descendants. This is indicated by the burst of aberrant body plans encountered among the earliest metazoans. After the early Cambrian, the tighter integration of the genotype produced increasingly more severe constraints on the capacity for producing structural innovation. However, the integration within a given body plan was still sufficiently loose to permit great variation,

TABLE 3.1   Estimated Time of Origin of Major Classes of Vertebrates

Vertebrate Class	Period	Time of Origin
Jawed fishes	Ordovician	450 million years ago
Lobe-finned fishes	Silurian	410 million years ago
Amphibians	Upper Devonian	370 million years ago
Reptiles	Upper Pennsylvanian	310 million years ago
Birds	Upper Triassic	225 million years ago
Mammals	Upper Triassic	225 million years ago

as shown by the radiation of the echinoderms, arthropods, and chordates, and the angiosperms among the plants.

Perhaps the most important conclusion from this evidence is that all the major subdivisions of the animal kingdom were already in existence in the Cambrian, more than 500 million years ago: the diploblasts (sponges and coelenterates), the triploblasts (Protostomia and Deuterostomia), and the major subdivisions of the Protostomia, the Ecdysozoa and the Spiralia (Table 3.1). No longer are there any enigmatic phyla whose relationship is totally unknown. Even the puzzling conodonts, so conspicuous as Paleozoic fossils, have now been unmasked as chordates. At the level of the classes there are still considerable uncertainties, particularly among the protists, whose phylogeny is still poorly understood. However, the overall picture of the classification and evolution of the metazoans (animals) is now reasonably well understood.

The Correct Evaluation of Characters

The validity of a classification largely depends on the proper evaluation of the characters on which it is based. Owing to their radial symmetry, Cuvier combined the coelenterates and the echinoderms in the higher taxon Radiata. However, it was soon shown how different the two radial taxa are in just about all their other characters, and it was realized that the radial symmetry of the echinoderms was due to convergent evolution of a basically bilateral body plan. Metamerism is characteristic for several phyla of animals, particularly the annelids, arthropods, and vertebrates. However, much evidence suggests that this character originated independently in the three mentioned groups. One must always make a careful test of homology when one encounters such similarities in otherwise rather different groups to determine whether or not their similarity is due to convergence. But convergent similarity may also develop when two unrelated taxa independently lose the same characteristic. For instance, it is very probable that nonsegmented groups such as the molluscs, Echiura, and Pogonophora descended from segmented ancestors.

Parallelophyly

An independent evolutionary acquisition of the same characters by unrelated groups may lead to the recognition of polyphyletic groups, like the "fishes" of Linnaeus, which included the whales. Such polyphyly must be distinguished from *parallelophyly*, the independent acquisition of the same character by several different descendants of a common ancestor (see Chapter 10). In the latter case, the ancestral genotype, shared by the descendants, produced the same phenotype independently. A striking illustration is the parallel evolution of the same trophic specializations in six lineages of endemic cichlid fishes in Lake Tanganyika in East Africa. Parallelophyly may be the explanation why the pelvis and legs of certain late Cretaceous bipedal dinosaurs are so strikingly similar to those structures in birds that are also bipedal. This explanation would be quite compatible with a Triassic derivation of birds from thecodont archosaurians, who were also the ancestors of the dinosaurs, and thus presumably had a rather similar genotype with the same morphological propensities (see the later section on The Origin of Birds, p. 65).

Phyletic Series

According to Darwinism there should be smooth continuity in the sequence of fossils in succeeding strata. Alas, as deplored by Darwin himself, the fossil record presents us with almost nothing but discontinuities: "The explanation [for these gaps] lies, as I believe, in the extreme imperfection of the geological record." Fortunately, since 1859 the fossil record has improved dramatically and we now have a large number of cases where the gradual change of a species into a derived species can be documented, step by step, and where even the gradual change of a genus into a derived genus can be followed. A particularly impressive example is the gradation of the therapsid reptiles through the cynodonts to the mammals. Several genera of cynodonts on this lineage already have certain mammalian characteristics and could be assigned to the mammals (see Fig. 2.1).

An even more complete gradation is presented by the evolution of the modern horse (see Fig. 2.3). A simple transitional genus *(Meryc-hippus)* gave rise to no less than nine new genera, one of which *(Dino-hippus)* gave rise to the modern horse *(Equus)*. A beautiful series of intermediate stages also exists between the mesonychid ungulates and their descendants, the whales (see Fig. 2.2). In most cases new species seem to have originated by budding in a peripherally isolated population, but such a localized population is not likely to be preserved in the fossil record. It suddenly appears on the scene and then continues essentially unchanged until it becomes extinct. This mode of phyletic evolution is particularly well documented for the bryozoan genus *Metaraptodos* (Cheetham 1987). Futuyma (1998) describes and illustrates numerous such cases of nearly complete phyletic series.

THE EVOLUTION OF PLANTS

The fossil record of the earliest plants is very poor. Fossils of mosses, generally considered the most primitive of the living land plants, have been found from the Devonian period, but surely they existed earlier and did not fossilize. They had apparently evolved from charo-phycean algae. Symbiotic fungi may have played an important role in the conquest of the inhospitable land. The first vascular plants were found in the Silurian. The dominant plants in the Paleozoic era (particularly the Carboniferous) were lycopods, ferns, and seed ferns. The Mesozoic was dominated by gymnosperms, particularly cycads and conifers, while the now dominant plants, the angiosperms, did not flourish until the Cretaceous, ca. 125 million years ago, even though they originated in the Triassic (Taylor and Taylor 1993). About 270,000 species of flowering plants have been described so far, placed in about 83 orders and 380 families. Through the application of a combination of morphological and molecular methods the relationship (phylogeny) of the orders of angiosperms is now reasonably well understood. The entire enormous radiation of the flowering plants occurred since the middle of the Cretaceous, coevolving with a similar radiation of insects.

THE ORIGIN OF THE VERTEBRATES

When we visit a large natural history museum, we find great halls showing the diversity of fish, amphibians, turtles, dinosaurs, birds, and mammals. The zoologists combine all of these creatures in the subphylum Vertebrata. These, in turn, are a subdivision of the phylum Chordata. Traditionally the other 30–35 phyla of animals were combined under the Invertebrata, even though this name concealed a highly diverse assortment of different kinds of animals. What are they and how did they evolve?

One group of protists, the choanoflagellates, gave rise to the sponges (Porifera), the simplest animals. From these rose the diploblastic coelenterates (Cnidaria, Ctenophora), which then gave rise to the triploblastic Bilateria, which soon split into the Protostomia and the Deuterostomia (see the earlier discussion). The Deuterostomia consist of four phyla: Echinodermata, Hemichordata, Urochordata, and Chordata. One of the earliest chordates, *Amphioxus*, is still surviving and shows approximately what our earliest ancestor looked like. Since it has gill slits and a dorsal notochord, *Amphioxus* is combined with the vertebrates in the phylum Chordata. *Amphioxus* was a filter feeder but it is inferred that the earliest vertebrates were predators. A closely related class of chordates are the extinct conodonts, which had an elaborate set of hard teeth that are copiously preserved in the fossil record.

The fossil record of the earliest vertebrates is rather poor. A recently found 530-my-old fossil from Yunnan (China) was described as a fish. The agnathan fishes (hagfishes and lampreys), traced back to

about 520 my ago, are still surviving, while the earliest toothed vertebrates (placoderms) are extinct. The inferred times of origin of the
later classes of vertebrates are given in Table 3.1.

The Origin of Birds

Whenever there is a large gap between the earliest undisputed ancestor of a new higher taxon and its later representatives, different authors may propose different branching points. This is well illustrated
by the origin of birds. The earliest undisputed bird fossil is *Archaeopteryx*, found in the upper Jurassic (145 million years ago). There are
two major proposals concerning the phylogeny of birds. According to
the thecodont theory, birds originated from archosaurian reptiles in
the late Triassic, maybe more than 200 million years ago. According
to the dinosaurian theory, birds originated from theropod dinosaurs
in the later Cretaceous (ca. 80 to 110 million years ago) (Fig. 3.5).
The main support for the dinosaurian theory is the extraordinary
similarity of the avian skeleton to that of certain bipedal dinosaurs,
particularly in the structure of the pelvis and the posterior extremities
(Fig. 3.6).

How can we determine which of the two conjectures is the correct
one? The most decisive refutation of the dinosaurian theory would be
a fossil bird or bird ancestor in the Triassic, let us say from a 220-
million-year-old stratum. Unfortunately there are no known bird fossils older than 150 million years. Actually one such fossil, *Protoavis*,
was described (Chatterjee 1997), but has not been examined by any
leading avian anatomist. Lacking a universally recognized fossil, the
proponents of the thecodont and dinosaurian derivations have cited
reasons why the proposal of their opponents cannot be valid. I have
listed in Box 3.3 the arguments of the thecodont origin supporters for
why a dinosaurian origin cannot be correct. But how can the extraordinary similarity in the walking apparatus of birds and dinosaurs be
explained? One possibility is to ascribe it to their similar bipedal locomotion and to parallelophyly. Both taxa were derived from the same
archosaurian phyletic lineage, although at very different times. The
thecodont ancestors of birds were close relatives of the ancestors of
the dinosaurs and can be assumed to have had a rather similar geno-

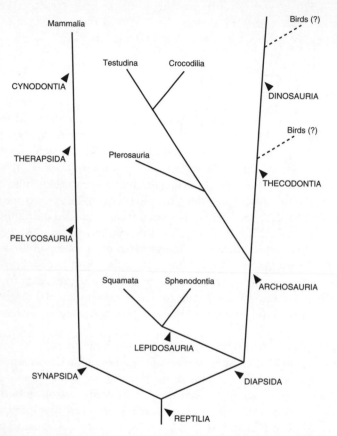

FIGURE 3.5

Highly schematic phylogeny of the Reptilia, showing the reptilian groups from which mammals and birds branched off. The geological timescale or degrees of similarity are not considered in this diagram.

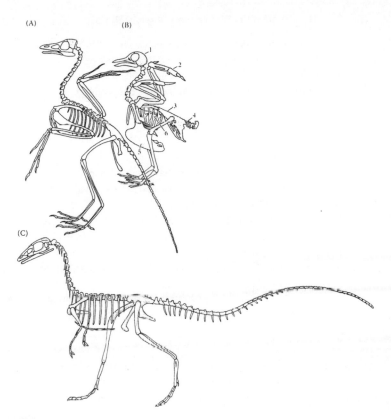

FIGURE 3.6

Similarities between birds and dinosaurs. A, *Archaeopteryx;* B, modern bird (pigeon); C, the theropod dinosaur *Compsognathus. Source*: Futuyma, Douglas J. (1998). *Evolutionary Biology* 3rd ed. Sinauer: Sunderland, MA.

Box 3.3   Refutation of the Dinosaurian Origin of Birds

1. Age—The dinosaurs structurally most similar to birds are very recent (80–110 million years ago), whereas *Archaeopteryx* is a great deal older (145 million years ago) and no birdlike dinosaurs are known from the lower Jurassic or Triassic that could qualify as ancestors of birds.

2. The three digits of the hand of the dinosaurs are 1,2,3, those of a bird are 2,3,4. It is quite impossible to derive the avian digits from those of dinosaurs.

3. Teeth—Theropods have recurved, flattened, serrated teeth, quite different from the simple peglike, waisted, nonserrated teeth of *Archaeopteryx* and other early birds.

4. The pectoral girdle and anterior extremities of the late theropod dinosaurs are much too small and weak to have served as the foundation of a powerful wing to lift an incipient bird from the ground. No factors are known that could have caused a sudden drastic growth of the anterior extremities.

5. The leading aerodynamic experts of bird flight claim that an origin of flight from the ground up is a near impossibility.

---

type as the dinosaurs. The shift to bipedal locomotion may have induced their similar genetic endowment to respond with a similar morphological construction as the bipedal birds. Only further fossils can settle the argument decisively.

CONCLUSIONS

Darwin's theory of common descent postulates that every group of organisms is derived from an ancestral group. An ancestral group, in turn, may have several descendant groups. In theory, it should be possible to establish the ancestry of every group of fossil or still living organism.

In 1859, when Darwin published *On the Origin of Species*, evolutionists were far from achieving this objective. None of the nearest relatives of any phyla was known. Yet T. H. Huxley was able to demonstrate that the class Aves (birds) undoubtedly had reptilian ancestry. The phylogenetic research of the following 140 years has resulted in establishing a seemingly well substantiated reconstruction of the major lines of descent. For example, the reptiles are derived from a group of amphibians and the amphibians from Rhipidistian fishes. When the ancestry leads us far back into the Precambrian, the recognition of groups such as the Deuterostomia and the Bilateria brings together related phyla even when some of the details of their descent have not yet been worked out.

What is most gratifying is that all findings are consistent with Darwin's theory of common descent. Together with molecular sequences, the fossil record, in spite of its many gaps, is the most irrefutable evidence for the occurrence of evolution. However, continuous fossil sequences are still the exception; the fossil record remains woefully inadequate. For instance, we have no fossil documentation of the human ancestry between 14 and 4.5 million years ago. The most recent coelacanth fossil is dated ca. 60 million years ago and, of course, everybody concluded that this group had become extinct that long ago, until two living species were discovered within the last 50 years. However, when even such an unexpected discovery has been made, it always fitted perfectly into the Darwinian framework.

II

HOW ARE
EVOLUTIONARY CHANGE
AND ADAPTEDNESS
EXPLAINED?

CHAPTER 4

HOW AND WHY DOES EVOLUTION TAKE PLACE?

The searching human mind is not satisfied merely to discover facts. We also want to know how things happen and why. And beginning with Darwin, evolutionists have devoted enormous ingenuity trying to answer these questions, and in the process have produced an abundance of answers. Depending on the kind of organisms they worked on (plants or animals, living or fossil) and depending on their philosophical background, they produced a plethora of theories, many of them in conflict with each other and with Darwin's original theories. After many years of controversy, in the 1940s a far-reaching consensus or evolutionary synthesis was achieved.

THE RETARDING INFLUENCE OF WIDELY HELD PHILOSOPHICAL VIEWS

Hindsight suggests that enough facts were available soon after 1859 to have permitted the universal acceptance of Darwin's theories, yet they were not universally adopted until about 80 years later. What could have been the reason for this long resistance? This is what historians have long asked themselves, but a satisfactory answer was not found until rather recently. The resistance, it was found, was due to the dominance of certain almost universally held philosophical ideas in the worldview of Darwin's opponents. A strict belief in the literal truth of every word in the Bible was one of them. Its power, however, was limited, as is shown by the rapid acceptance (except by creation-ists) of Darwin's theory of common descent. However, several other ideologies in conflict with Darwin's theories were essentialism and finalism.

To refute these erroneous ideas, Darwin introduced four new con-cepts—population thinking, natural selection, chance, and history (time)—all of which were largely or entirely missing from the philos-ophy of science in the middle of the nineteenth century. Thus Darwin not only refuted the opposing ideologies but he also introduced the new concepts that eventually became the foundation of the philoso-phy of biology as it developed after 1950. It is quite impossible to un-derstand the nature of the post-Darwinian controversies unless one understands the nature of the ideologies opposed to Darwinism. A presentation of their basic tenets is therefore necessary.

Typological Thinking (Essentialism)

Essentialism was the almost universally held worldview from the an-cients until Darwin's time. Founded by the Pythagoreans and Plato, essentialism taught that all seemingly variable phenomena of nature could be sorted into classes. Each class is characterized by its defini-tion (its essence). This essence is constant (invariable) and sharply de-marcated against all other such essences. For instance, the Pythagore-ans said, a triangle is always a triangle, no matter what shape it has, and is not connected by intermediates with quadrangles or any other geometric figures. The class of trees is defined by a trunk and a leafy crown. A horse is characterized (defined) by its high teeth and a foot with a single toe. In Christian religious belief, each kind, each type, each species is believed to have been separately created and all now living members of a species are believed to be the descendants of the first pair created by God. The essence or definition of a class (type) is completely constant; it is the same today as it was on the day of the Creation. Essentialism was adhered to not only by Christians, but also by most agnostic philosophers. All seeming variation among the members of a class was considered "accidental" and irrelevant. A species was considered by the essentialist to be such a class and was referred to by the philosophers as a *natural kind*.

The early pre-Darwinian evolutionists (including Lamarck) adopted a weakened version of strict essentialism by allowing a gradual change (transformation) of the type over time. At any given time, however, the type was still considered to be more or less invariable.

Population Thinking

Darwin made a radical break with the typological tradition of essentialism by initiating an entirely new way of thinking. What we find among living organisms, he said, are not constant classes (types), but variable populations. Every species is composed of numerous local populations. Within a population, in contrast to a class, every individual is uniquely different from every other individual. This is true even for the human species with its six billion individuals. Darwin's new way of thinking, being based on the study of populations, is now referred to as *population thinking*. This approach was congenial to most naturalists, who in their systematic studies had discovered that species of animals and plants showed as much (and sometimes far more) variation and uniqueness as the human species. The gradual replacement of essentialism by population thinking led to long-lasting controversies in evolutionary biology. All saltational theories of evolution are based on essentialism, whereas population thinking favors the acceptance of gradualism. Population thinking is one of the most important concepts in biology: It is the foundation of modern evolutionary theory and one of the basic constituents of the philosophy of biology (see below).

Finalism

Another non-Darwinian ideology in the nineteenth and early twentieth centuries was *finalism*, the belief that the living world has the propensity to move toward "ever greater perfection." Those who adopted finalism assumed that evolution moved necessarily from lower to higher, from primitive to advanced, from simple to complex, from imperfect to perfect. They postulated the existence of some built-in force, because, they said, how else can one explain the gradual evolution from the lowest bacteria up to orchids, giant trees, butter-flies, apes, and man? This belief in finalism goes at least as far back as Aristotle, who recognized it as one of the causes, indeed the final cause. For many years after 1859, finalism was still accepted by a large proportion of evolutionists (see below), but never by Darwin. Darwin emphatically rejected such obscure forces. Instead, he fully accepted Newton's credo that everything in the world is controlled by purely mechanical (physical-chemical) forces. However, Darwin introduced a historical perspective into science, which was absent from Newton's explanatory framework. Almost invariably one must invoke historical antecedents in the explanation of evolutionary phenomena.

Such opposing ideologies as essentialism and finalism prevented the immediate acceptance of Darwin's explanation of the how and why of evolution. And so for the first 80 years after the publication of *On the Origin of Species*, Darwin's theory of variational evolution had to battle with three other major evolutionary theories attempting to explain evolution. Since these theories still receive occasional support even today, it is important to understand their claims and weaknesses. Indeed, a discussion of the deficiencies of the theories competing with Darwinism contributes toward a better understanding of the strength of the theory of variational evolution.

WHAT EVOLVES?
••••••••••••••••••••

Almost everything in the inanimate universe is also evolving, that is, it is changing in a distinctly directional sequence. But what is it that evolves in the living world? Species surely evolve, and so do all combi-nations of species in the Linnaean hierarchy—genera, families, orders, and all higher taxa up to the totality of the living world. But what about lower levels? Do individuals evolve? Certainly not in any genetic sense. To be sure, our phenotype changes in the course of our life, but our genotype remains essentially the same from birth to death. Then what is the lowest level of living organization to evolve? It is the popu-lation. And the population turns out to be the most important site of evolution. *Evolution is best understood as the genetic turnover of the indi-viduals of every population from generation to generation.*

To complete the precise characterization of evolution in sexually reproducing species, it is necessary to define the evolutionary population. A local population *(deme)* consists of the community of potentially interbreeding individuals of a species at a given locality (see Chapter 5). Curiously, the concept of population, as here described, was unknown before 1859; even Darwin was inconsistent in its application. Everybody else tended to think in terms of types.

Once we acknowledge the presence of the various opposing ideologies in Darwin's period, it is easier to understand why so many different theories of evolution were promoted, and it is also easier to see the weaknesses that led to their ultimate downfall.

Box 4.1   Theories of Evolution Based on Essentialism vs. Population Thinking

A. Based on Essentialism

1. Transmutationism: Evolution occurs through the production of new species or types, owing to a mutation or saltation.

2. Transformationism: Evolution occurs through the gradual transformation of an existing species or type into a new one, either

   a. by the direct influence of the environment or by use and disuse of the existing phenotype, or
   b. by an intrinsic drive toward a definite goal, particularly toward greater perfection, and
   c. through an inheritance of acquired characters.

B. Based on Population Thinking

3. Variational (Darwinian) Evolution: A population or species changes through the continuous production of new genetic variation and through the elimination of most members of each generation, because they are less successful either in the process of the nonrandom elimination of individuals or in the process of sexual selection (i.e., they have less reproductive success).

THREE THEORIES OF EVOLUTION BASED ON ESSENTIALISM
....

Transmutationism

If one believes that all phenomena in the world are manifestations of underlying constant types, as stated in the philosophy of essentialism, change can happen only through the origin of new types. Since a type (essence) cannot evolve gradually (types are considered to be constant!), a new type can originate only through an instantaneous "mutation" or *saltation* of an existing type, which thereby gives rise to a new class or type. For the supporters of this view, often called saltationists, the world is full of discontinuities. The transmutationist postulates that a mutation results in the sudden origin of a new kind of individual. This individual, together with its offspring and their descendants, represents a new species.

Rudiments of *transmutationism* go all the way back to the Greek philosophers, but it was also adopted in the eighteenth century by the French philosopher Maupertuis, and after 1859 not only by many opponents of Darwinism, but even by some of Darwin's friends, including T. H. Huxley. Although *saltationism* was vigorously criticized by Weismann and other Darwinians, it remained popular for almost 100 years. Several leading geneticists in the early 1900s, the so-called Mendelians (De Vries, Bateson, Johannsen), were such saltationists. The last prominent defenses of this theory were published as late as the mid-twentieth century (Goldschmidt 1940; Willis 1940; Schindewolf 1950).

Saltationism was popular for such a long time not only because it was consistent with the philosophy of essentialism, but also because it seemed to be consistent with the observations of the naturalists. All species in a local fauna and flora seemed to be sharply demarcated against each other, and the appearance (as well as the disappearance) of new species in the fossil record seemed invariably to be an instantaneous event. Wherever one looked in nature, one found discontinuities, and not the gradual changes postulated by Darwin. One could not refute saltationism until one could explain why there are so many discontinuities ("gaps") where one expected gradations (intermediates). Before this could be answered, however, considerable advances in species-level taxonomy were needed and these did not happen until well into the twentieth century.

Many different observations and arguments led to the final refutation of transmutationism. First was the realization that a species is not a type that can mutate to a new type, but rather comprises many populations. Not all the individuals in a population can have the same mutation simultaneously. Therefore a new species could not originate instantaneously. Those who postulated that transmutation happens through the origin of a single newly mutated individual were up against other formidable difficulties. The genotype of an individual is a harmonious, well-balanced system, brought together through millions of years and fine-tuned by natural selection in every generation. Since it was known that potential mutations at most gene loci have deleterious or lethal effects, how could a massive shake-up of an entire genotype by a major mutation possibly produce a viable individual? Only an incredibly rare individual (called a "hopeful monster" by Goldschmidt) would have any chance for survival and success, whereas the great mass of such macromutants would be failures. But where are all of these millions of failures resulting from such a macromutational process? They have never been found because, as is now quite obvious, such a postulated macromutational process does not occur.

The terms gradual and discontinuous have different applications and can lead to misunderstandings if these are not distinguished. When Darwin proclaimed gradualness and continuity, he had the seeming gap between taxa in mind. Even though a gap may now exist between two species, it did not necessarily originate by a saltation. As we now know, there never was a "taxic discontinuity," because the two species were connected with their common ancestor by a continuous series of intermediate populations. On the other hand, individuals in a single population may differ by visibly different characters—blue eyes vs. brown eyes, two molars vs. three molars, or even more conspicuous differences. Such "phenotypic discontinuity" within a population characterizes all cases of polymorphism. A successful mutation with a large phenotypic effect can be gradually incorporated into a population as long as it is able to pass through a period of polymorphism in which it coexists with the previous phenotype, until it has completely displaced the original gene. Admittedly it is sometimes difficult to understand how a certain new phenotype was thus acquired. The cheek pouch of the pocket gopher is an example.

Darwin never tired of emphasizing that most evolutionary changes take place by very small steps. But this is not true for all of them. There are chromosomal phenomena, particularly in plants (polyploidy) and in certain animal groups (parthenogenetic species hybrids), that in a single step can produce a new species (see Chapter 9). But these are marginal occurrences that do not invalidate the overwhelming prevalence of gradual populational evolution. Still, it must be remembered that there is a considerable range in the size of the mutations that lead to evolutionary change.

Transformationism

In the eighteenth century the evidence for evolution became so widespread and impressive that it was no longer compatible with classical typology. The theory of essentialism was therefore somewhat relaxed: The type could become gradually "transformed" in the course of time, although it was still essentially invariable at any given moment in time. Even though a type could change, it remained the same object. Evolution of a species, it was said, was like the development of a zygote from the fertilized egg to the adult. Indeed, the term evolution was first used by the Swiss philosopher Bonnet for the preformation theory of individual development. In Germany, ontogeny and evolution were both called up to the twentieth century by the same term, *Entwicklung*. This concept of gradual evolution is called *transformationism*. It is applied to any theory that is based on the gradual change of an object or its essence. All seemingly evolutionary processes in the inanimate world fall into this category. Examples are the change of a star from one type (white, yellow, red, blue) to another type; or the gradual rise of a mountain range owing to tectonic forces and its subsequent destruction by erosion. Two attributes are characteristic for transformationism: the change of a specific object and the gradual continuity of the change.

Darwin's friend and mentor, the geologist Charles Lyell, was a staunch supporter of this kind of gradualism, which he called *uniformitarianism*. For Lyell, all changes in nature, and particularly those in the history of the Earth, had been gradual. There were no discontinuities, no sudden jumps ("saltations") or instant mutations. Lyell's influence was a major factor in Darwin's adoption of gradualism, although Darwin's populational gradualism was something entirely different from Lyell's uniformitarianism.

As far as the living world is concerned, one can distinguish two drastically different transformational theories of evolution: that based on environmental influences and that based on the strive for perfection.

Transformation Owing to Environmental Influences.   According to this theory—often but not quite correctly also called the Lamarckian theory   evolution is caused by the gradual change of organisms owing either to "use and disuse" of a structure or other trait or to the direct influence of the environment on the genetic material. This theory assumes that the genetic material is "soft" and that it can be molded by environmental influences, and that these changes can then be transmitted to future generations by an "inheritance of acquired characters." This theory is based on a belief in soft inheritance.

The most often cited case of an inheritance of acquired characters is the long neck of the giraffe. According to Lamarck, it was stretched in every generation by each giraffe trying to reach the highest accessible tree branches for feeding, and this lengthening of the neck was inherited by the next generation. Likewise, if a structure is not used, such as the eye in cave animals, it gradually withers. Not only use and disuse could produce inheritable changes, it was claimed, but also the direct influence of the environment. Prior to Darwin, it was widely believed that the Negroes had black skin because they had been exposed for thousands of generations to the tanning effects of the tropical sun. Many characteristics of organisms were attributed to such a direct influence of the environment.

Transformationism was undoubtedly the most widely adopted evolutionary theory from 1859 until the evolutionary synthesis of the 1940s. Even though natural selection was for Darwin the principal factor in evolution, he also accepted the idea of soft inheritance, perhaps as a source of variation. In the presynthesis period, most naturalists, following Darwin, also accepted both natural selection and soft inheritance.

Lamarckism explained gradualism and was widely accepted by the opponents of transmutationism. However, all experiments that tried to demonstrate its validity were unsuccessful. Mendelian genetics, by proving the constancy of genes, completely contradicted soft inheritance. Finally, it was shown by molecular biology that no information can be transmitted from the proteins of the body to the nucleic acids of the germ cells, in other words, that an inheritance of acquired characters does not take place. This is the so-called "central dogma" of molecular biology.

Transformation Owing to a Strive for Perfection (Orthogenesis). This theory (or set of theories) is based on a belief in cosmic teleology (finalism). According to this belief, the living world has the propensity to move toward ever greater perfection. Theories of this type, held by authors like Eimer, Berg, Bergson, Osborn, and many other evolutionists, are referred to as orthogenetic or autogenetic theories. They postulated that types (essences) were steadily improved by an intrinsic drive, and that evolution was believed to take place not by the origin of new types, but by the transformation of existing types. These theories were abandoned when no mechanism could be found to drive such trends. Furthermore, such drives, if they existed, should result in "rectilinear" (straight) evolutionary lineages, but the paleontologists showed that all evolutionary trends sooner or later change their direction or may even reverse themselves. Finally, one can explain linear trends as the product of natural selection. Indeed there is no evidence whatsoever to support any belief in cosmic teleology.

The refutation of the existence of final causes was of fundamental importance for philosophy, because it was one of the causes postulated by Aristotle and had occupied an important place in the teaching of most philosophers. Kant's acceptance of teleology had greatly affected the thinking of German evolutionists in the nineteenth century.

All three endeavors to explain this world and its changes (evolution) on the basis of typological thinking (essentialism) were thus a failure. It was necessary to adopt an entirely different approach, and this was found by Charles Darwin and Alfred Russel Wallace.

CHAPTER 5

··

VARIATIONAL EVOLUTION

Variation played no role in transmutationism nor in either of the two kinds of transformationism. All three theories were strictly based on essentialism. "Evolution" takes place in transmutationism through the origin of a new essence and in the transformationist theories through a gradual change of the essence.

VARIATION AND POPULATION THINKING
···

Darwin showed that one simply could not understand evolution as long as one accepted essentialism. Species and populations are not types, they are not essentialistically defined classes, but rather are biopopulations composed of genetically unique individuals. This revolutionary insight required an equally revolutionary explanatory theory of evolution: Darwin's theory of variation and selection. Two sources of evidence led Darwin to this new concept. One was the empirical study of variable natural populations (particularly during his study of the barnacles), and the other was the observation by animal and plant breeders that no two individuals of their herds or breeding stocks were identical. These individuals were not members of essentialistic classes, and, as we now know, all individuals in a sexual population are genetically unique.

Apparently, most people find it difficult to grasp the significance of this uniqueness. Let them remember that no two individuals among the 6 billion humans are identical, not even so-called identical (monozygous) twins. An understanding of the fundamental difference between a class of essentially identical objects and a biopopulation of unique individuals is the foundation of so-called "population thinking," one of the most important concepts of modern biology.

> The assumptions of population thinking are diametrically opposed to those of the typologist. The populationist stresses the uniqueness of everything in the organic world. What is true for the human species—that no two individuals are alike—is equally true for all other species of animals and plants. Indeed, even the same individual changes continuously throughout its lifetime and when placed into different environments. All organisms and organic phenomena are composed of unique features and can be described collectively only in statistical terms. Individuals, or any kind of organic entities, form populations of which we can determine the arithmetic mean and the statistics of variation. Averages are merely statistical abstractions, only the individuals of which the populations are composed have reality. The ultimate conclusions of the population thinker and of the typologist are precisely the opposite. For the typologist, the type *(eidos)* is real and the variation an illusion, while for the populationist the type (average) is an abstraction and only the variation is real. No two ways of looking at nature could be more different. (Mayr 1959)

Darwin's Variational Evolution

It was Darwin who introduced this new way of thinking into science. His basic insight was that the living world consists not of invariable essences (Platonian classes), but of highly variable populations. And it is the change of populations of organisms that is designated as evolution. Thus, evolution is the turnover of the individuals of every population from generation to generation.

When Darwin, in 1837, became an evolutionist (see Chapter 2), he asked himself, How can the process of evolution be explained? Could he adopt one of the already proposed explanations? He realized eventually that neither transmutationism nor transformationism nor any other theory based on essentialism would do. And he was right. All essentialistic theories of organic evolution are badly flawed, as was convincingly established during the post-Darwinian controversies.

Darwin had to develop an entirely new kind of explanation that accounted for the abundance of variation in nature. This led him to his theory of natural selection, which was based on population thinking (see Chapter 6). The same theory was found independently by Alfred Russel Wallace.

Although Darwin published *On the Origin of Species* in 1859 (actually Wallace and Darwin published a first statement in 1858), the explanatory theory of variational evolution was not universally adopted until ca. 80 years later. It is a theory based on the variability of populations. There were two sets of practitioners who had already appreciated this variability for a long time, the taxonomists and the animal and plant breeders, and Darwin had close connections to both of them.

When sorting out the collections he had made on the voyage of the *Beagle*, Darwin encountered the same question again and again: Are some slightly different specimens merely variants within a population or are they different species? Indeed, in the 1840s when he wrote his monographs on the classification of the barnacles, Darwin came to the conclusion that no two specimens in a collection from a single population were exactly identical. They all were as uniquely different from each other as are human individuals. And the animal and plant breeders, with whom Darwin was associated since his Cambridge student days, told him the same. They always knew which individuals in their herds they should select as the breeding stock for the next generation. Individuality made this possible.

Since the terms "transmutationism" and "transformationism" are not suitable for this new theory, Darwin's theory of evolution through natural selection is best referred to as the theory of *variational evolution*. According to this theory, an enormous amount of genetic variation is produced in every generation, but only a few individuals of the vast number of offspring will survive to produce the next generation. The theory postulates that those individuals with the highest probability of surviving and reproducing successfully are the ones best adapted, owing to their possession of a particular combination of attributes. Since these attributes are largely determined by genes, the genotypes of these individuals will be favored during the process of selection. As a consequence of the continuous survival of individuals (phenotypes) with genotypes best able to cope with the changes of the environment, there will be a continuing change in the genetic composition of every population. This unequal survival of individuals is due in part to competition among the new recombinant genotypes within the population, and in part to chance processes affecting the frequency of genes. The resulting change of a population is called evolution. Since all changes take place in populations of genetically unique individuals, evolution is by necessity a gradual and continuous process.

Darwin's Theories of Evolution

Darwin's views on evolution are often referred to as The Darwinian Theory. Actually they consist of a number of different theories that are best understood when clearly distinguished from each other. The most important of Darwin's theories of evolution are discussed below (see Box 5.1). That they are indeed five independent theories is documented by the fact that the leading "Darwinians" among Darwin's contemporaries accepted some and rejected others (see Box 5.2).

First   Two of these five theories, evolution as such and the theory of common descent, were widely accepted by biologists within a few years of the publication of the *Origin*. This represented *the first Darwinian revolution*. The acceptance of man as a primate in the animal kingdom was a particularly revolutionary step. Three other theories, gradualism, speciation, and natural selection, were strongly resisted and were not generally accepted until the evolutionary synthesis. This was *the second Darwinian revolution*. The Darwinism proposed by Weismann

2nd

Box 5.1   Darwin's Five Major Theories of Evolution

1. The nonconstancy of species (the basic theory of evolution)
2. The descent of all organisms from common ancestors (branching evolution)
3. The gradualness of evolution (no saltations, no discontinuities)
4. The multiplication of species (the origin of diversity)
5. Natural selection

Box 5.2   Rejection of Some of Darwin's Theories by Early Evolutionists

The following table shows the composition of the evolutionary theories of various evolutionists. All of these authors accepted a fifth theory, that of evolution as opposed to a constant, unchanging world. They differed in accepting or rejecting some of Darwin's four other evolutionary theories.

	Common Descent	Gradualness	Populational Speciation	Natural Selection
Lamarck	No	Yes	No	No
Darwin	Yes	Yes	Yes	Yes
Haeckel	Yes	Yes	?	In part
Neo-Lamarckians	Yes	Yes	Yes	No
T. H. Huxley	Yes	No	No	No
de Vries	Yes	No	No	No
T. H. Morgan	Yes	No	No	Unimportant

---

and Wallace, in which an inheritance of acquired characters is rejected, was named *Neodarwinism* by George John Romanes. The Darwinism accepted since the evolutionary synthesis is best simply called *Darwinism*, because in most crucial aspects it agrees with the original Darwinism of 1859, while the belief in an inheritance of acquired characters is by now totally obsolete.

Darwin's theory of gradualism fitted well into the thinking of the transformationists, but the resistance of the saltationists was sufficiently great that the universal acceptance of the gradualness of evolution had to await the evolutionary synthesis. Darwin's concept of gradualness, however, was of an entirely different nature from that of the transformationists. Their gradualness was due to the gradual change of an essential type, whereas Darwinian gradualism is due to the gradual restructuring of populations. This makes it quite clear why Darwinian evolution, being a populational phenomenon, must always be gradual (see Chapter 4). A Darwinian must be able to show that every seeming case of saltation or discontinuous evolution can be explained as being caused by a gradual restructuring of populations.

VARIATION

The availability of variation is the indispensable prerequisite of evolution, and the study of the nature of variation is therefore a most important part of the study of evolution. Variation, the uniqueness of every individual, is, as we said, characteristic of every sexually reproducing species. To be sure, at first sight all the individuals of a species of snail or butterfly or fish might seem identical, but a closer study of these individuals will reveal all sorts of differences in size, proportions, color pattern, scaling, bristles, and whatever characteristic one studies. Further studies have shown that variability affects not only visible characters, but also physiological traits, patterns of behavior, aspects of ecology (e.g., adaptation to climatic conditions), and molecular patterns, all of this reinforcing the conclusion that in one way or another every individual is unique. And it is this always available variability that makes the process of natural selection possible.

Although the variability of the phenotype was appreciated by naturalists as far back as Darwin's day, the early geneticists treated the genotype as rather uniform. When the studies by the population geneticists from the 1920s to the 1960s revealed the presence of a great deal of cryptic variation, this was questioned by some of the classical authors. Yet not even the most enthusiastic Darwinians suspected the amount of genetic variation in populations that was eventually revealed by the methods of molecular genetics. Not only was it discovered that much of the DNA consists of noncoding DNA ("junk"), but it was also found that many, perhaps the majority of alleles are "neutral," that is, their mutation does not affect the fitness of the phenotype (see below). As a result, it is now realized that seemingly identical phenotypes may conceal considerable variation at the level of the gene.

Polymorphism

Sometimes variation falls into definite classes, a phenomenon referred to as *polymorphism*. In the human species we have polymorphisms for eye color, hair color, straightness or curliness of the hair, different blood groups, and many other of the genetic variants of our species. The study of polymorphisms has greatly contributed to our understanding of the strength and direction of natural selection, as well as the causal factors underlying variation. Two outstanding studies are those on the color polymorphism of banded snails *(Cepaea)* by Cain and Sheppard and on chromosome arrangements in *Drosophila* by Dobzhansky. In most cases it is unknown what is responsible for the maintenance of polymorphism in a population over long periods. A balance of selection pressures is usually assumed, but it may be reinforced by some superiority of the heterozygous carriers that favors the retention of the rarer gene in the population. In a highly diverse environment, phenotypic diversity may be selected, as in the case of the banded snails.

THE SOURCE OF VARIABILITY

What is the source of this variability? Where does it come from? How is it maintained from generation to generation? This is what puzzled Darwin all of his life, but in spite of all his efforts he never found the answer. An understanding of the nature of this variability was finally made possible, after 1900, by advancements in genetics and molecular biology. One can never fully understand the process of evolution unless one has an understanding of the basic facts of inheritance, which explain variation. Therefore the study of genetics is an integral part of the study of evolution. But only the heritable part of variation plays a role in evolution.

Genotype and Phenotype

As early as the 1880s it was recognized by perceptive biologists that the genetic material (germ plasm) was something different from the body of an organism (soma), and this distinction was satisfied when the early Mendelians introduced the terms *genotype* and *phenotype*. But the prevailing opinion at that time was that the genetic material consisted of proteins like those that make up the body. It came as a real shock when Avery demonstrated in 1944 that the genetic material consisted of nucleic acids. The terminological distinction between an organism and its genes now acquired a new meaning. The genetic material itself is the genome (haploid) or the genotype (diploid), which controls the production of the body of an organism and all of its attributes, the phenotype. This phenotype is the result of the interaction of the genotype with the environment during development. The amplitude of variation of the phenotype produced by a given genotype under different environmental conditions is called its *norm of reaction*. For instance, a given plant may grow to be larger and more luxuriant under favorable conditions of fertilizing and watering than without these environmental factors. Leaves of the Water Buttercup (*Ranunculus flabellaris*) produced under water are feathery and very different from the broadened leaves on the branches above water (see Fig. 6.3). As we shall see, it is the phenotype that is exposed to natural selection, and not individual genes directly.

It has been heatedly argued in the past whether a particular property of an organism was due to "nature" (its genes) or "nurture" (its environment). All research in the last 100 years indicates that most characteristics of an organism are affected by both factors. This is particularly true for characters that are controlled by multiple genes. There are two sources of variation in a sexually reproducing population, superimposed on each other: the variation of the genotype (because in a sexual species no two individuals are genetically identical) and the variation of the phenotype (because each genotype has its own norm of reaction). Different norms of reaction may react rather differently to the same environmental conditions.

THE GENETICS OF VARIATION

We owe our understanding of variation to the branch of biology called genetics, which is devoted to the study of the nature of inheritance. This science has grown, since its founding in 1900, into one of the largest biological disciplines and is extremely rich in fact and theory. Even textbooks restricted to evolutionary genetics may run to more than 300 pages. I am forced in this work on evolution to limit my treatment to an analysis of the basic principles of genetics, leaving a more detailed treatment to the special texts of the field. Maynard Smith (1989) and Hartl and Jones (1999) are recommended for fuller detail. For a beginner, the genetics chapter of any biology text, such as that of Campbell (1999), will be helpful, or the more extensive genetics chapters in the evolution books of Futuyma (1998), Ridley (1996), and Strickberger (1996). Fortunately, an understanding of the basic principles of genetics necessary for an understanding of evolution does not require all the detail offered in these books. I feel that it is sufficient to understand a limited number of basic principles, but these must be understood thoroughly. The seventeen principles listed here would seem to be the most important ones.

Seventeen Principles of Inheritance

1. The genetic material is constant ("hard"); it cannot be changed by the environment or by use and disuse of the phenotype. The inheritance of constant genetic material is called hard inheritance. Genes cannot be modified by the environment. Properties acquired by the proteins of the phenotype cannot be transmitted to the nucleic acids of the germ cells. There is no inheritance of acquired characters.
2. The genetic material, as was discovered by Avery in 1944, consists of DNA (deoxyribose nucleic acid) molecules (in some viruses also RNA). The DNA molecule has a double-helix structure, as discovered by Watson and Crick in 1953 (Fig. 5.1).
3. The DNA contains the information that permits the production of the proteins that (together with lipids and other molecules) make up the phenotype of every organism. It controls the assemblage of amino acids that are converted into proteins with the help of cellular structures and mechanisms.
4. In the eukaryotes most DNA is located in the nucleus of every cell and is organized into a number of longitudinal bodies called *chromosomes* (Fig. 5.2). (Small amounts of DNA and RNA occur also in cellular organelles, such as mitochondria and chloroplasts.)
5. Sexually reproducing organisms are normally *diploid*, that is, they have two homologous sets of chromosomes, one inherited from the male parent and the other from the female parent.
6. Both male and female gametes have only one chromosome set, and so are *haploid*. When the egg is fertilized, diploidy is restored to the newly formed organism (*zygote*), because the

chromosomes of the two parents do not fuse but remain discrete (see principle 7). This is why Mendelian inheritance is called particulate.

7. During the *fertilization* of an egg by a spermatozoon, the chromosomes of the male parent (containing the paternal genes) do not fuse or blend with the chromosomes of the female parent (containing the maternal genes) but rather coexist in the fertilized egg (zygote). The genetic material is thus handed unchanged from generation to generation, except for an occasional mutation (see principle 11).

8. Characteristics of organisms are controlled by genes, which are located on the chromosomes.

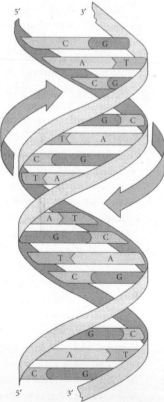

FIGURE 5.1

The well-known double helix of DNA. The base pairs, always one purine and one pyrimidine, are the horizontal "steps" of the helical staircase. *Source*: Futuyma, Douglas J. (1998). *Evolutionary Biology* 3rd ed. Sinauer: Sunderland, MA.

9. A gene is a sequence of nucleic acid base pairs that encodes a program with a specific function.

10. On the whole, the nuclei of all cells of the body contain the same genes.

11. Although a gene is normally constant from generation to generation, it has the capacity to "mutate" occasionally into a different form. Such a newly mutated gene (mutant) will again be constant, unless another new mutation occurs.

12. The totality of the genes of an individual constitute its genotype.

13. Each gene has a number of different forms, called *alleles*, which are responsible for most of the differences among the different individuals of a population (Fig. 5.3).

14. A diploid organism has a pair of each gene, one from the male parent and one from the female parent. If these two genes are the same allele, the organism is called *homozygous* for this gene; if they belong to different alleles, the organism is called *heterozygous*.

15. When, in a heterozygote, only one of the two alleles is expressed in the phenotype, it is called the dominant allele; the other allele is called recessive.

16. A gene has a complex structure, consisting of exons, introns, and flanking sequences (Fig. 5.4).

17. There are several different kinds of genes, some of which control the actions of other genes (see below).

Age of Genes.   Perhaps the most unexpected result of modern molecular studies of the genome was the discovery of the great age of many genes. The sequence of base pairs is often so conservative that one can determine that a certain mammalian gene is also part of the genome of the fruit fly *Drosophila* or the nematode *Caenorhabditis*. Indeed it seems possible to trace some genes all the way from animals or plants to bacteria. This fact is particularly important in the study of disease genes. For instance, one can treat a mouse with an inserted human disease gene with all sorts of drugs to test their curative capacity. It is also of great potential for the application of genetic engineering. Even where such practical applications are not possible, a comparison of the same gene in different kinds of organisms usually makes an important contribution to our understanding of gene functions.

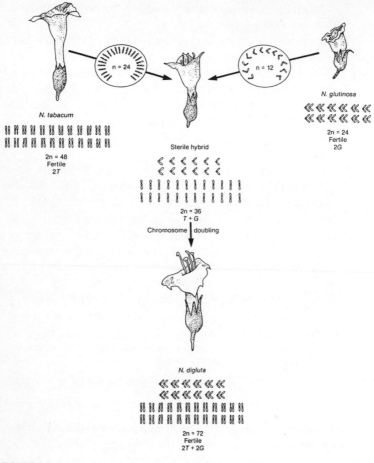

N. tabacum

2n = 48
Fertile
2T

N. glutinosa

2n = 24
Fertile
2G

Sterile hybrid

2n = 36
T + G
Chromosome doubling

N. digluta

2n = 72
Fertile
2T + 2G

FIGURE 5.2

Origin of polyploidy. A cross between two species of plants often produces a sterile hybrid. A doubling of the chromosome number may, in certain crosses, produce a fertile allopolyploid species with the double chromosome number. *Source*: Strickberger, Monroe, W., *Evolution*, 1990, Jones and Bartlett, Publishers, Sudbury, MA. www.jbpub.com. Reprinted with permission.

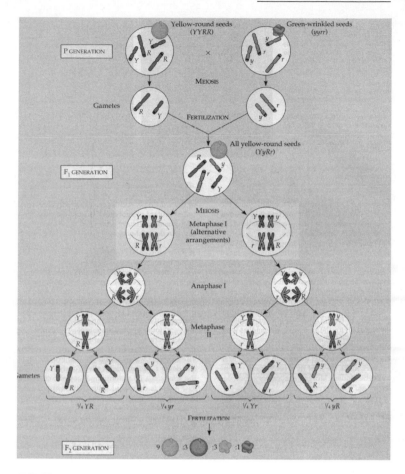

FIGURE 5.3

A gene may have different versions called alleles. In one of Mendel's crosses he used gene Y with two alleles Y (dominant; yellow seeds) and y (recessive; green seeds), and gene R with two alleles R (dominant; round seeds) and r (recessive; wrinkled seeds). Crosses with these two sets of alleles gave the results shown in this figure. *Source*: Figure 15.1, p. 262 from *Biology* 5th edition, by Neil A. Campbell, Jane B. Reece, and Lawrence G. Mitchell. Copyright © 1999 by Benjamin/Cummings, an imprint of Addison Wesley Longman, Inc. Reprinted by permission of Pearson Education, Inc.

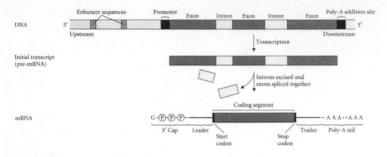

FIGURE 5.4

Structure of a eukaryotic gene, with its exons, introns, and flanking sequences. *Source*: Futuyma, Douglas J. (1998). *Evolutionary Biology* 3rd ed. Sinauer: Sunderland, MA.

GENETIC TURNOVER IN A POPULATION

According to the Hardy-Weinberg equation, the gene contents of a population would remain the same from generation to generation if it were not for a number of processes that may lead to the loss of existing genes or to the acquisition of new genes. These processes are responsible for evolution (see Box 5.3).

Seven such processes are of particular evolutionary importance: selection, mutation, gene flow, genetic drift, biased variation, movable elements, and nonrandom mating. Selection will be treated in Chapter 6; the other six will now be discussed.

Mutation

The use of the term *mutation* in biology has had a checkered history (Mayr 1963: 168–178). Prior to 1910 it was used for any drastic change of the type, particularly when such change instantaneously produced a new species. Morgan (1910) restricted mutation to a spontaneous change of the genotype, more precisely to a sudden change of a gene. Gene mutations are due to errors of replication during cell division. Even though the replication of the DNA molecules during cell division and gamete formation is remarkably accurate, occasional er-

---

Box 5.3   The Hardy-Weinberg Principle

In the early years of genetics there was much confusion about what determined allele frequencies in a population. However, in 1908 G. H. Hardy in England and W. R.Weinberg in Germany showed mathematically that allele frequencies in populations would remain constant, generation after generation, if certain processes did not occur that would lead to the loss of existing genes or the acquisition of new genes. They expressed this as a mathematical formula, which is a reapplication of a mathematical law, the binomial expansion. Being a strictly mathematical solution, it is not a *biological* law.

Let us illustrate this by an example. Let us assume a gene is represented by two alleles in a population, A1 and A2. The frequency of A1 is p and that of A2 is q, with $p + q = 1$. The following frequencies of gametes will be present at reproduction and will produce the frequencies of genotypes as follows:

		Sperm	
		A1 (p)	A2 (q)
Eggs	A1 (p)	A1A1 (p^2)	A1A2 (pq)
	A2 (q)	A1A2 (pq)	A2A2 (q^2)

The binomial expansion $(p + q) (p + q) = p^2 + 2pq + q^2$ will be maintained generation after generation unless there is an addition or loss of genes (see text).

---

rors do occur. The replacement of a base pair by a different one is called a gene mutation. However, there are also larger changes of the genotype, such as polyploidy or changes of the gene arrangement, as occurs in chromosomal inversion. These are referred to as chromosomal mutations. Any changes on the pathway from the DNA of the gene (messenger RNA, ribosomes) to the amino acids or polypeptides of the phenotype are also classified as mutations. Mutations may also be caused by the insertion of a transposable element in the chromosome. Any mutation that induces changes in the phenotype will either be favored or discriminated against by natural selection.

According to their evolutionary significance, three kinds of mutations can be distinguished: beneficial, neutral, or deleterious. Individuals with genotypes that contain a beneficial new mutation will be favored by natural selection. However, since almost all conceivable beneficial mutations of a population in a stable environment have already been selected in the recent past, the occurrence of new beneficial mutations is rather rare. Mutations that do not affect the fitness of the phenotype, so-called neutral mutations, are frequent. Their evolutionary role will be discussed below. Finally, deleterious mutations will be selected against and will be eliminated in due time. If they are recessive, they may survive in a population in heterozygous condition. If they result in immediate elimination, they are called lethal. The selective value of a gene may vary depending on its interaction with the remainder of the genotype.

Even though all new genes are produced by mutation, most of the phenotypic variation in natural populations that is available for selection is the product of recombination (see below). Before the role of selection was fully understood, it was believed by many evolutionists that some evolutionary changes were due to "mutation pressure." This is a misconception. The frequency of a gene in a population is in the long run determined by natural selection and stochastic processes, and not by the frequency of mutation.

Gene Flow

The gene content of every local population, except the most isolated ones, is strongly affected by the immigration and emigration of genes to and from other populations of the species. This exchange of genes among neighboring populations is called *gene flow*. Gene flow is a conservative factor that prevents the divergence of only partially isolated populations and it is a major reason for the stability of widespread species and for the stasis of populous species. The amount of gene flow differs from population to population and from species to species. Highly sedentary *(philopatric)* species have little gene flow, whereas those with a strong dispersal tendency may be almost *panmictic*.

It is important to realize that *dispersal* propensity seems to be highly variable among individuals of a given population. Indeed, there may be a pronounced polymorphism in that respect. Certain individuals of a population may be highly philopatric, reproducing very close to their place of birth; others may disperse relatively short distances; and a few individuals may move far away from their birthplace, sometimes up to several hundred kilometers. These latter individuals, of course, are of the greatest evolutionary significance. Most of them probably will be unsuccessful, not being optimally adapted for their new location, yet these long-distance colonizers may establish *founder populations* and discover suitable locations well beyond the current species range.

Some species are so successful at dispersal that they have a cosmopolitan distribution, as is true for species with spores or animal species that have wind-dispersed eggs, as found among the tardigrades and certain crustaceans. However, even relatively short dispersal can efficiently counteract any tendency for a progressive divergence of local populations. Gene flow is an extremely conservative factor in evolution.

Genetic Drift

In a small population alleles may be lost simply through errors of sampling (stochastic processes); this is known as *genetic drift.* Indeed, such a random loss of alleles may occur even in rather large populations. This is usually of no consequence in widespread species, because such locally lost genes will be quickly replaced by gene flow in subsequent generations. However, small founder populations, beyond the periphery of the range of a species, may have a rather unbalanced sampling of the gene pool of the parental population. This may facilitate a restructuring of the genotype of such populations (see below).

Biased Variation

Some genes (so far known only in a few species) affect the segregation of alleles during meiosis in a heterozygote such that the allele of one parental chromosome goes to the gametes in more than half of the instances. If this allele controls an unfit phenotype, it will be selected against. Only rarely is such biased variation sufficiently strong to override the eliminating power of selection.

Transposable Elements

Transposable elements (TEs) are DNA sequences ("genes") that do not occupy fixed sites on a chromosome but can move to a new site on the same or a different chromosome. There are various kinds of TEs with various effects. When inserted in a new location on a chromosome, they may cause a mutation on an adjacent gene. They often produce short DNA sequences that replicate frequently. One of these sequences, called Alu, is highly repeated with more than 500,000 copies in an individual of many mammalian species. It constitutes about 5 percent of the human genome. No selectively valuable contributions are known for any of the TEs. Rather they often seem deleterious, but natural selection seems unable to eliminate them. Confer a genetics textbook for a detailed treatment of the manifold manifestations of transposable elements.

Nonrandom Mating

In all species with sexual selection there may be a preference by one sexual partner for a particular phenotype of its mate. This leads to a nonrandom favoring of certain genotypes.

Some cases of *sympatric speciation* are best explained as products of nonrandom mating. In certain groups of fishes, particularly cichlid fishes, females seem to mate preferentially with males that prefer a certain subniche. If, for instance, in a lake in which at first species A occupies and feeds in both the benthic and the limnetic (open-water) zones and a group of females preferentially mates with benthic males, these females will select simultaneously for any visible markings that characterize males that prefer to feed in the benthic niche. Feeding and mating are no longer random and gradually two subpopulations evolve, the members of which preferentially feed and mate either benthically or limnetically. In due time the two subpopulations may evolve into two fully isolated sympatric species. In most groups of fishes this mode of sympatric speciation apparently does not occur. The same kind of process may lead to sympatric speciation in host-specific insects if mating preferentially takes place on the plant for which both mates have the same preference.

UNIPARENTAL REPRODUCTION AND EVOLUTION

Success in Darwinian evolution depends on the continuous availability of large amounts of variation. The greatest part of this phenotypic variation tends to be produced by recombination of the parental chromosomes, that is, by sexual reproduction, a process invented by the eukaryotes. However, large numbers of organisms do not have sexual reproduction; these organisms use uniparental reproduction. How do they manage to produce the variation needed to keep up with the changes in their environment?

In most forms of uniparental ("asexual") reproduction, the offspring are genetically identical with the parent. A lineage produced by such reproduction is called a *clone*. How does a clone acquire new genetic variation? In higher organisms this is accomplished normally only through mutation. Any new mutation gives rise to a new mini-clone. If it is a successful mutation, the new clone will prosper and through acquiring additional mutations will gradually diverge from the parental clone. Eventually, as happened in the bdelloid rotifers, the differences between the most successful clones may become as great as those between different species of sexually reproducing species. Unsuccessful clones become extinct, and this is how the gaps between the "species" are created in asexual higher taxa.

Prokaryotes reproduce asexually. They acquire genetic variation by mutation and by unilateral exchange of genes with other clones. But as soon as sexuality had been, so to speak, "invented," asexuality became relatively rare among the eukaryotes. Above the level of the genus there are only three higher taxa of animals that consist exclusively of uniparentally reproducing clones. Strict asexuality is rare in plants but common in some groups of fungi.

Prokaryotes reproduce asexually. All prokaryote individuals are, so to speak, of the same sex. Sexual reproduction is unknown among them. Yet sexual reproduction is now the almost universal mode of eukaryote reproduction. Every case of uniparental reproduction found in higher animals and plants is obviously a secondary (derived) condition, usually being restricted to a single species in a genus or to an isolated genus. There are only a few cases of entire families of animals being parthenogenetic (see below). It is rather obvious that in animals uniparental reproduction has been invented again and again, but the asexual clones always became extinct after a relatively short time.

Sexual versus Asexual Reproduction

What does the relative rarity of asexuality among the eukaryotes suggest? It leads to the inference that uniparental reproduction, where it is now found in higher organisms, is not primitive, but a derived condition. It has evolved independently, again and again, in unrelated groups, but soon becomes extinct. No matter what the selective advantage of sexual reproduction is, that it must have an advantage is clearly indicated by the consistent lack of success of asexuality.

And yet asexual reproduction would seem at first sight to be far more productive than sexual reproduction. Let us take a population with two kinds of females, both having the same number of 100 offspring, reduced in every generation to two survivors. Females A reproduce sexually and of its offspring 50 are males and 50 are females. Females B reproduce asexually and produce 100 females. A very simple calculation shows that in a short time the population would consist almost exclusively of the asexual B females.

An asexually reproducing female that can produce fertile eggs (parthenogenesis) does not "waste" any gametes on the production of males, and thus has twice the fertility of a sexual individual that produces both kinds of gametes. Why then does natural selection not favor parthenogenesis, the ability of females to produce eggs that do not require fertilization by males?

Since the 1880s evolutionists have argued over the selective advantage of sexual reproduction. So far, no clear-cut winner has emerged from this controversy. As often occurs in such controversies, plural answers may be the right answer. In other words, sexual reproduction has several advantages and combined they outweigh the seeming numerical advantage of asexuality. We must first grasp the entire process of sexual reproduction before we can understand why sexuality, in spite of its lower fertility, is in the long run more successful than asexual reproduction.

MEIOSIS AND RECOMBINATION
••

It took more than 100 years of study to achieve a full understanding of the meaning and the process of sexual reproduction. Darwin searched unsuccessfully all his life for the source of genetic variation. It required knowledge of the process of gamete formation and the difference between genotype and phenotype and their roles in natural selection, as well as an understanding of populational variation.

August Weismann and a group of cytologists found the answer. They showed that in sexual reproduction, gamete formation is preceded by two special cell divisions (see Box 5.4). During the first division, homologous maternal and paternal chromosomes attach themselves tightly to each other and then may break at one or several places. The broken chromosomes exchange parts with each other so that they now consist of a mixture of paternal and maternal chromo-

---

Box 5.4   Meiosis

Meiosis is the name for the two consecutive cell divisions that precede the formation of the haploid gametes. At the first division, sister chromatids of homologous chromosomes attach to each other. They may break at points of overlap in a process called *crossing-over*. A broken chromatid may join the broken end of the sister chromatid and become a composite new chromosome. In the ensuing second cell division, called the reduction division, homologous chromosomes go randomly to the opposite poles, thus producing entirely new chromosomic sets. Thus at two consecutive steps an entirely new recombination of the parental genotypes is produced through crossing-over and the random movement of homologous chromosomes to opposite poles.

The gametes (spermatozoa and eggs) produced during meiosis are haploid, but diploidy is restored by fertilization. Consult a biology textbook for further details of this complex process.

Crossing over

some pieces. This process is called *crossing-over.* Each new chromosome is an entirely new combination of maternal and paternal genes. In the second cell division preceding the formation of the gametes, the chromosomes do *not* divide, but one of each pair of homologous chromosomes goes randomly to one daughter cell and the other chromosome to the other daughter cell. As a result of this "reduction division" the "haploid" number of chromosomes in each gamete is half that of the "diploid" chromosome number of the zygote produced by the fertilized egg. This sequence of two cell divisions preceding gamete formation is called *meiosis.*

Two processes during meiosis achieve a drastic *recombination* of the parental genotypes: (1) crossing-over during the first division and (2) the random movement of homologous chromosomes to different daughter cells (gametes) during the reduction division. The result is the production of completely new combinations of the parental genes, all of them uniquely different genotypes. These, in turn, produce unique phenotypes, providing unlimited new material for the process of natural selection.

No matter what the selective advantage of sexual reproduction may be, that it does have such an advantage in animals is clearly indicated by the consistent failure of all attempts to return to asexuality. Obligatory asexuality is not found among higher plants, but agamospermy, seed production without fertilization, is widespread (Grant 1981). Uniparental reproduction, however, is more frequent than sexual reproduction in certain protists, fungi, and some groups of nonvascular plants. It is the exclusive mode of reproduction in the prokaryotes, in which unidirectional gene transfer provides genetic variation.

WHY IS THE PRODUCTION OF SUCH HIGHLY VARIABLE GENOTYPES SO FAVORED BY SELECTION?

Occasional asexual reproduction is of wide occurrence in the animal kingdom (but absent in birds and mammals). In almost every case it is restricted to a single species in an otherwise sexual genus or to an asexual genus. Only three higher taxa of animals (above the level of the genus) consist exclusively of uniparentally reproducing clones (bdelloid rotifers and some ostracods and mites). It is quite obvious

3 animals to produce clones

that species have experimented with "buying" doubled fertility by abandoning sexuality, but the asexual clones die out sooner or later.

For more than a century evolutionists have speculated about the nature of sexuality's powerful advantage, but until now no unanimity has been reached. Surely when a population suddenly encounters an extremely adverse situation, the more genetically diverse it is, the greater is the chance that it contains genotypes that can better cope with the environmental demands, compared to a uniform clone or a group of closely related clones.

A considerable number of solutions have been proposed for the mechanism by which sexuality (recombination) is favored by selection. They all have in common a greater survival of beneficial mutations and a faster elimination of deleterious mutations in sexual populations than in asexual ones. Pathogens (new diseases), for example, are best coped with by the origin of new resistant genotypes. The genotype, consisting of nucleic acids, is not directly exposed to natural selection, but is translated during the development of the fertilized egg into the proteins and other constituents of the phenotype (see Chapter 6). The phenotype is the result of the interaction of the environment with the genotype.

The process of sexual reproduction makes far more new phenotypes available for natural selection than does mutation or any other process. It is the major source of the variation found in populations of sexual species. This capacity for the production of large amounts of variation would seem to be the major selective advantage of sexual reproduction (see the special section "The Evolution of Sex," *Science* 281(1988): 1979–2008). It is this capacity for recombination that gives sexual reproduction its enormous evolutionary importance.

Recombination

A member of a population in a sexually reproducing species mates with another member of its population and they produce in their offspring an entirely new recombination of the parental genes. The phrase "gene pool" for the genes found in a population is somewhat misleading. The genes are not independently swimming in a "pool," but are linearly arranged on the chromosomes, each individual in a sexually reproducing diploid species carrying on its chromosomes one haploid set of paternal and one haploid set of maternal genes. This is the Sutton-Boveri theory, first put forth at the turn of the twentieth century, and later confirmed by T. H. Morgan. This diploid combination of the parental genetic material is called the genotype. Each individual is a unique combination of the two sets of parental genes, and it is the phenotype, the product of the genotype (the recombined set of genes), that is ordinarily the actual target of selection (see below). Recombination in a population is the major source of the phenotypic variation available for effective natural selection.

Lateral Transfer

There is no sexual reproduction in the prokaryotes and thus no replenishment of genetic variation by recombination. Instead, genetic variation in bacteria is renewed by a process called unidirectional lateral transfer, in which a bacterium attaches itself to another one and transfers some of its genes. There is little information on the types of genes transferred by this process. It is probably limited to certain classes of genes, since the major types of bacteria, such as gram-negative, gram-positive, and cyanobacteria, are not fused by this process. Even the archaebacteria exchange genes with other families of bacteria.

What happened to lateral transfer after the origin of sexual reproduction? Until the 1940s, it was assumed that this process had disappeared among sexually reproducing organisms. However, Barbara McClintock then discovered in maize the transposons, or genes that move from their position on one chromosome to another chromosome. This is such a new and unexpected discovery that it is not yet clear whether the phenomenon is widespread. There are also nucleic acid entities (e.g., plasmids) that are largely independent of the chromosomes. These genetic elements are of particular importance among the asexually reproducing prokaryotes. Whenever they affect the phenotype, they are subject to natural selection.

Gene Interaction

How the phenotype is produced by the action of the genes is the subject of physiological or developmental genetics. For the sake of simplicity, it was traditionally assumed that each gene acted independently of all others. This is not correct. Indeed, there are numerous interactions among the genes. Many genes, for example, may affect simultaneously several aspects of the phenotype. Such genes are called *pleiotropic*. Pleiotropy is most conspicuously demonstrated by deleterious genes, like the genes for sickle cell anemia (see Box 6.3), cystic fibrosis, and similar mutations, which affect some basic tissue activity that manifests itself in numerous different organs. On the other hand, a particular aspect of the phenotype may be affected by several different genes. Such inheritance is called *polygenic* inheritance. Pleiotropy and polygeny contribute to the cohesion of the genotype; the multiple interactions of genes is referred to as *epistasis*.

These interactions of genes are the least understood properties of the genotype. They will be referred to again in later chapters in connection with phenomena such as evolutionary stasis, bursts of evolutionary change, and mosaic evolution. The so-called "cohesion of the genotype" is one of the aspects of these interactions (see below). The study of the structure of the genotype is the most challenging of all future tasks of evolutionary biology.

Genome Size

If the production of new genes would parallel evolutionary advance, one would expect that the organisms that are highest on the phylogenetic tree would have the largest genome. Up to a point this is indeed true. Genome size is measured in terms of the number of base pairs, although for practical reasons the units are megabases (1,000 base pairs, abbreviated Mb). The genome of humans is about 3500 Mb. In a bacterium it may be only 4 Mb. Very large figures were found in salamanders and lungfishes. An equally great variation was found in plants.

Why should there be such enormous variation and, in particular, such great differences among closely related organisms? The answer is that there are two kinds of DNA, that active in development (coding genes) and that not active (noncoding DNA) (see Box 5.5). The great differences in the Mb numbers are almost completely due to the presence of smaller or greater amounts of noncoding genes, of-

---

Box 5.5   Noncoding DNA

A remarkably high proportion of the DNA in the chromosomes seems
not to perform an obvious function such as coding for RNAs and proteins. Such DNA, sometimes probably incorrectly referred to as "junk,"
is estimated for humans to be as much as 97 percent of the total DNA.
This portion of our genome includes introns, repetitive sequences such
as microsatellite DNA, and various kinds of "interspersed elements"
such as Alu sequences. There is a widespread belief among Darwinians
that such apparently unnecessary DNA would have been eliminated
long ago by natural selection if it did not have some, as of yet undiscov-
ered, function. Indeed the introns have a recognized function, to keep
the exons apart prior to the activation of a gene (translation of the DNA
message into proteins). During the translation process the introns are
excised prior to the translation of a gene into proteins. Introns also contain many regulatory elements (DNA motifs that serve as binding sites
for transcription regulation genes) and are thought to enhance eukaryotic genetic complexity via alternative splicing through both *cis-* and
trans-acting elements.

---

ten referred to as "junk." There are numerous mechanisms by which
noncoding genes are produced and multiplied, particularly by retro-
transposable elements. There are also mechanisms by which junk
DNA is eliminated, and different species differ in the efficiency of
their elimination mechanisms. Research on the factors that control
genome size still has a long way to go before full understanding is
achieved. The size of the active genome is not only much smaller,
but also far less variable than these numbers suggest.

THE ORIGIN OF NEW GENES

A bacterium has about 1,000 genes. A human has perhaps 30,000
functional genes. Where did all these new genes come from? They
originate by duplication, with the duplicated gene inserted in tandem

in the genome next to the sister gene. Such a new gene is called a *par-alogous* gene. At first, it will have the same function as its sister gene. However, it will usually evolve by having its own mutations and in due time it may acquire functions that differ from those of its sister gene. The original gene, however, will also evolve, and such direct descendants of the original gene are called *orthologous* genes. In homology studies only orthologous genes may be compared.

Additions to the genome come not only by the duplication of single genes, but sometimes through the duplication of groups of genes, whole chromosomes, and entire chromosome sets. For instance, a special mechanism, involving the kinetochores, can lead to a duplication of chromosome sets in certain orders of mammals, leading to highly variable chromosome numbers in these orders. Lateral transfer is another way for addition to the genome.

Kinds of Genes

Molecular biology has discovered that there are many kinds of genes. Some directly control the production of organic material (via enzymes) and others control the activity of the material that produces genes. No mutation in 8,000 of the 12,000 genes of the *Drosophila* genome seems to have an effect on the phenotype. Changes in these genes have been referred to as neutral evolution (see below).

Genes that do not code for proteins have long been considered to be "junk." However, they may play an important but not yet understood role in the regulation of other genes. The explanation of the role of the noncoding DNA may provide the solution to some of the open questions about the structure of the genotype. There are several different kinds of noncoding genetic material, including introns, pseudogenes, and highly repetitive DNA (Li 1997). At least some noncoding DNA definitely has a function: introns keep the exons separate. What is particularly difficult to understand is the great amount of noncoding DNA. According to some estimates, 95 percent of the human DNA is "junk." A Darwinian finds it difficult to believe that selection would not have been able to get rid of it if it was indeed totally useless. After all, the production of this DNA is expensive.

Homeobox Genes, Regulatory Genes

All living animals belong to a limited number of basic designs: radially symmetrical, bilaterally symmetrical, segmented (metamerical), and characteristic subdivisions of these basic patterns. The great German morphologists have referred to such a basic design as *Bauplan*, which was translated into English (not quite correctly) as "body plan." In German, the syllable *plan* in *Bauplan* means "map" or "blueprint" —not something that someone had planned. It is not a metaphysical concept.

Until a few years ago it was a complete riddle how a set of genes could determine what in the development of the zygote should become the anterior or the posterior end of the embryo, or the dorsal or the ventral side, and in a metameric organism which segment should bear what appendages. However, developmental genetics has now provided many explanations. In addition to the substrate-producing "structural" genes, there are regulatory genes that produce proteins able to determine front or rear, ventral or dorsal, and so on (*Hox* genes), or the construction of special organs, like the eye (*pax* gene). Sponges have only a single *Hox* gene, arthropods have 8, and mammals have 4 *Hox* clusters with 38 genes. Mice and flies share 6 *Hox* genes, which the common ancestor of Protostomia and Deuterostomia already must have had (see Box 5.6).

Everything indicates that the basic regulatory systems are very ancient and were later coopted for additional functions when these were acquired (Erwin et al. 1997). Such specialized developmental genes are largely independent of the action of other genes and permit the independent development of different parts and structures of the developing embryo. For example, the development of wings in a bat can take place with minimal disturbance of the other developmental pathways. This explains why so-called mosaic evolution is such a widespread phenomenon.

THE NATURE OF VARIATION
······································

In Darwin's day, the nature of variation in populations was not yet understood. This understanding was possible only following develop-

Box 5.6   *Hox* Genes

Developmental as well as evolutionary biologists aim to better understand the evolution of complexity and the origin of morphological novelties in evolution through the analyses of the expression patterns of *Hox* genes during the ontogeny of organisms. It is suspected that these genes might play a pivotal role in specifying regional identity in body plans. *Hox* genes are arranged in genomic clusters and code for a class of transcription factors (genes that control the expression of other genes), and, importantly, their expression takes place in a spatially and temporally colinear fashion. Anterior genes in the *Hox* gene clusters are expressed earlier in development and more anteriorly in the embryo, whereas posterior genes are switched on later in development and in more distal portions of the body.

It has been suggested that increasing complexity of body plans during evolution might be causally correlated with increasing complexity of the *Hox* gene complexes. Invertebrates have only a single *Hox* gene cluster, and the common ancestor of all chordates probably also had only a single set of 13 *Hox* genes. During the evolution of chordates from relatively simply and rather segmentally organized cephalochordates like *Amphioxus* to more complex organisms like mice and humans, who have four *Hox* gene complexes, the single ancestral cluster probably duplicated twice for a complete set of 52 *Hox* genes in four clusters. These duplications from one to two and then to four clusters (A–D) occurred either as individual chromosome duplications, rather than through tandem duplications, because the clusters are each on different chromosomes, or through the duplications of entire genomes. Later in evolution, individual *Hox* genes on these clusters were lost on particular evolutionary lineages, yet mice and humans have the same set of 39 *Hox* genes distributed on four *Hox* clusters. None of these clusters retained their original set of 13 genes and each contains a unique combination of genes.

Differences in the gene content and expression patterns of *Hox* genes are assumed to be at least partially responsible for the different body plans that differentiate phyla of animals. The function of many *Hox* genes is paradoxically often extremely conserved in evolution, permitting remarkable experiments that demonstrated that *Hox* genes from, for example, *Amphioxus* can rescue the function of homologous genes in mice that had been experimentally removed from these mice. It remains an open question how new body plans are specified and evolved in light of, or in spite of, the remarkably conserved genomic architecture of *Hox* gene clusters and their highly conserved function in evolution.

ments in the late nineteenth and twentieth centuries. What Darwin did know as a naturalist, taxonomist, and student of natural populations was that variation in natural populations seemed to be virtually inexhaustible. It provides abundant material for natural selection in all organisms, at least in sexually reproducing species of animals and plants. The visible characteristics of an organism, its phenotype, are due to instruction during development by their genes and by the genotype interacting with the environment.

The Impact of the Molecular Revolution

Although the basic principles of inheritance were worked out between 1900 and the 1930s, the real understanding of the nature of inheritance was achieved only through the molecular revolution. It began in 1944 (Avery et al.) when it was established that the genetic material consisted not of proteins but of nucleic acids. In 1953 Watson and Crick discovered the structure of DNA, and after this one major discovery followed the other, culminating in the discovery of the genetic code by Nirenberg in 1961 (Kay 2000). Finally, every step in the translation of the genetic information in the course of the developing organism was understood in principle. Unexpectedly, the basic Darwinian concepts of variation and selection were not affected in any way. Not even the replacement of proteins by nucleic acids as the carriers of genetic information required a change in the evolutionary theory. On the contrary, an understanding of the nature of genetic variation greatly strengthened Darwinism, for it confirmed the finding of the geneticists that an inheritance of acquired characters is impossible.

Molecular biology's greatest contribution to evolutionary biology was the creation of the field of developmental genetics. Developmental biology, which had so long resisted the evolutionary synthesis, now adopted Darwinian thinking and analyzed the functional role of the genotype. This led to the discovery of regulatory genes (*hox, pax*, etc.) and thus vastly enlarged our understanding of the evolutionary aspects of development.

Evolutionary
Developmental Biology

One of the most important discoveries of molecular genetics was that some genes are very old. This means that the same gene (essentially the same sequence of base pairs) is found in organisms that are only very distantly related, say in *Drosophila* and mammals. A second discovery was that certain genes, often referred to as regulatory genes, control such basic developmental processes as the determination of anterior vs. posterior or of dorsal vs. ventral. These findings shed considerable light not only on previously completely puzzling developmental processes, but also on the causation of fundamental events (branching points) in phylogeny.

Scientists had always assumed that the same gene, no matter where found, always had the same phenotypic effect. But developmental geneticists have now shown that this is not necessarily so. The same gene may have rather different expressions in annelids (polychaetes) and arthropods (crustaceans). Selection seems to be able to recruit genes in new developmental processes that previously had seemed to have other functions.

It had been shown by morphological-phylogenetic research that photoreceptor organs (eyes) had developed at least 40 times independently during the evolution of animal diversity. A developmental geneticist, however, showed that all animals with eyes have the same regulatory gene, *Pax 6*, which organizes the construction of the eye. It was therefore at first concluded that all eyes were derived from a single ancestral eye with the *Pax 6* gene. But then the geneticist also found *Pax 6* in species without eyes, and proposed that they must have descended from ancestors with eyes. However, this scenario turned out to be quite improbable and the wide distribution of *Pax 6* required a different explanation. It is now believed that *Pax 6*, even before the origin of eyes, had an unknown function in eyeless organisms, and was subsequently recruited for its role as an eye organizer.

CONCLUSIONS
••••••••••••••••••••

It is shown in this chapter that Darwin, by making biopopulations the foundation of his evolutionary theorizing, rather than Platonic types, had found an entirely novel solution for the explanation of evolution. He postulated that the inexhaustible genetic variation of a population, together with selection (elimination), is the key to evolutionary change. To understand how this is implemented one must understand inheritance, and much of this chapter is therefore devoted to an explanation of the genetic basis of variation. The genetic material is constant and does not permit an inheritance of acquired characteristics. The genotype, interacting with the environment, produces the phenotype during development. Mutations continually replenish the variability of the gene pool. However, the variation of the phenotypes that provide the material for selection is produced by recombination in meiosis, a process of restructuring and reassorting the chromosomes.

· ·

CHAPTER 6

It was not until the 1930s that evolutionists fully appreciated (as was shown in Chapters 2–4) that none of the explanations of evolution based on essentialism was valid. Curiously, the correct explanation had been found by Darwin 100 years earlier in 1838, although it was not published until 1858/1859. This was the concept of natural selection. The dramatic novelty of the Darwin–Wallace theory was that it was based on population thinking instead of essentialism. Alas, essentialism was the dominant way of thinking of the period and it took several generations before natural selection was universally adopted. However, population thinking had a compelling logic as soon as one adopted it.

Natural selection as proposed by Darwin and Wallace was a most novel and daring theory. It was based on five observations (facts) and three inferences (see Box 6.1). When one refers to populations in discussing natural selection, one ordinarily has sexually reproducing species in mind, yet it also takes place among the clones of asexual organisms.

The theory of natural selection proposed by Darwin and Wallace became the cornerstone of the modern interpretation of evolution. It was a truly revolutionary concept, having never before been suggested by any philosopher, and only rather casually referred to by two of Darwin's contemporaries (William Charles Wells and P. Matthews). Even today many people have difficulty understanding how this principle works. Yet when population thinking is employed, it would seem to be simplicity itself. However, because the concept met

---

Box 6.1 Darwin's Explanatory Model of Natural Selection

Fact 1. Every population has such high fertility that its size would increase exponentially if not constrained. (Source: Paley and Malthus)

Fact 2. The size of populations, except for temporary annual fluctuations, remains stable over time (observed steady-state stability). (Source: universal observation)

Fact 3. The resources available to every species are limited. (Source: observation, reinforced by Malthus)

   Inference 1. There is intense competition (struggle for existence) among the members of a species. (Source: Malthus)

Fact 4. No two individuals of a population are exactly the same (population thinking). (Source: animal breeders and taxonomists)

   Inference 2. Individuals of a population differ from each other in the probability of survival (i.e., natural selection). (Source: Darwin)

Fact 5. Many of the differences among the individuals of a population are, at least in part, heritable. (Source: animal breeders)

   Inference 3. Natural selection, continued over many generations, results in evolution. (Source: Darwin)

---

strong resistance from long-established traditions and ideologies as the exclusive direction-giving factor to evolution, it remained a minority view from 1859 to the 1930s.

To better appreciate the difficulty of understanding natural selection, one must take a closer look at this process. We must ask Darwinian questions. For instance, what happens in a given population through time? How does a population change from generation to generation? What is responsible for these changes and how do they affect the populations of a species?

POPULATION
··················

Wherever a species occurs it is represented by a local population. Owing to unequal survival and reproductive success of its individuals, there is a continuing genetic turnover in each population as a result of chance and natural selection. Neighboring populations grade into each other if the habitat is continuous. However, favorable habitats are often discontinuous, resulting in a "patchy" distribution of the populations. Even greater breaks in the continuity of the populations occur where geographical barriers (mountains, water, unsuitable vegetation) inhibit dispersal. Along the border of a species' range, populations are often rather isolated.

An understanding of the nature of populations is of the utmost importance for an understanding of evolution, because all evolution, and particularly selection, takes place in biopopulations. All aspects of populations are, therefore, of interest to the evolutionist. A local population is sometimes called a *deme*. It may be defined as the community of potentially interbreeding individuals at a given locality.

As we have seen, the concept of natural selection is based on observations of the natural world. Every species produces vastly more offspring than can survive from generation to generation. All the individuals of a population differ genetically from each other. They are exposed to the adversity of the environment, and almost all of them perish or fail to reproduce. Only a few of them, on the average two per set of parents, survive and reproduce. However, these survivors are not a random sample of the population; their survival was aided by the possession of certain attributes that favor survival.

NATURAL SELECTION IS REALLY
A PROCESS OF ELIMINATION

The conclusion that these favored individuals had been selected to survive requires an answer to the question, Who does the selecting? In the case of artificial selection, it is indeed the animal or plant breeder who selects certain superior individuals to serve as the breeding stock of the next generation. But, strictly speaking, there is no such agent involved in natural selection. What Darwin called natural selection is actually a process of elimination. The progenitors of the next generation are those individuals among their parents' offspring who survived owing to luck or the possession of characteristics that made them particularly well adapted for the prevailing environmental conditions. All their siblings were eliminated by the process of natural selection.

Herbert Spencer, when saying that natural selection is nothing but "the survival of the fittest," was indeed quite right. Natural selection is a process of elimination, and Darwin adopted Spencer's metaphor in his later work. However, his opponents claimed that it was a tautology, a circular statement, by defining "the fittest" as those who survive, but this is a misleading claim. Actually, survival is not a property of an organism but only an indication of the existence of certain survival-favoring attributes. To be fit means to possess certain properties that increase the probability of survival. This interpretation is equally applicable to the "nonrandom survival" definition of natural selection. Not all individuals have an equal probability for survival because the individuals that have properties making survival more probable are a restricted nonrandom component of the population.

Do selection and elimination differ in their evolutionary consequences? This question never seems to have been raised in the evolutionary literature. A process of selection would have a concrete objective, the determination of the "best" or "fittest" phenotype. Only relatively few individuals in a given generation would qualify and survive the selection procedure. That small sample would be able to preserve only a small amount of the whole variance of the parental population. Such survival selection would be highly restrained.

By contrast, a mere elimination of the less fit might permit the survival of a rather larger number of individuals because they have no obvious deficiencies in fitness. Such an enlarged sample would provide, for instance, the needed material for the exercise of sexual selection. This also explains why survival is so uneven from season to season. The percentage of the less fit in a population would depend on the severity of each year's environmental conditions.

The larger the sample of the population that successfully passes the nonrandom process of elimination of the unfit, the more the success of the survivors will depend on chance factors and on selection for reproductive success.

The metaphor of selection pressure is frequently used by evolutionists to indicate the severity of selection. Even though it is a picturesque expression, this term, borrowed from the physical sciences, could be misunderstood, for there is no force or pressure connected with natural selection that corresponds to the use of the term in the physical sciences.

SELECTION IS A TWO-STEP PROCESS

Almost all of those who opposed natural selection failed to realize that it is a two-step process. Not realizing this, some opponents have called selection a process of chance and accident, while others have called it deterministic. The truth is that natural selection is both. This becomes obvious as soon as one considers the two steps of the selection process separately.

At the first step, consisting of all the processes leading to the production of a new zygote (including meiosis, gamete formation, and fertilization), new variation is produced. Chance rules supreme at this step, except that the nature of the changes at a given gene locus is strongly constrained (see Box 6.2).

At the second step, that of selection (elimination), the "goodness" of the new individual is constantly tested, from the larval (or embryonic) stage until adulthood and its period of reproduction. Those individuals who are most efficient in coping with the challenges of the environment and in competing with other members of their population and with those of other species will have the best chance to survive until the age of reproduction and to reproduce successfully. Numerous experiments and observations have revealed that certain individuals with particular attributes are clearly superior to others

---

Box 6.2   The Two Steps of Natural Selection

Step One: The Production of Variation
Mutation of the zygote from its origin (fertilization) to death; meiosis, with recombination through crossing-over at the first division, and random movement of homologous chromosomes during the second (reduction) division; any random aspects of mate choice and fertilization.

Step Two: Nonrandom Aspects of Survival and Reproduction
Superior success of certain phenotypes throughout their life cycle (survival selection); nonrandom mate choice, and all other factors that enhance the reproductive success of certain phenotypes (sexual selection). At the second step much random elimination occurs simultaneously.

during this process of elimination. They are the ones that are "fittest to survive." On the average, only two individuals of the abundant offspring of a set of parents will survive and become the progenitors of the next generation. This second step is a mixture of chance and determination. Clearly, those individuals with characteristics providing the greatest adaptedness to the current circumstances have the greatest probability of survival. However, there are also many chance elimination factors, so that there is no pure determination even at this step. Everything is somewhat probabilistic. Natural catastrophes, like floods, hurricanes, volcanic eruptions, lightning, and blizzards, may kill otherwise highly fit individuals. Furthermore, in small populations superior genes may be lost owing to sampling errors.

The fundamental difference between the first and the second steps of natural selection should now be clear. At the first step, that of the production of genetic variation, everything is a matter of chance. However, chance plays a much smaller role at the second step, that of differential survival and reproduction, where the "survival of the fittest" is to a large extent determined by genetically based characteristics. To claim that natural selection is entirely a chance process reveals total misunderstanding.

IS SELECTION A MATTER OF CHANCE?

Natural selection, unexpectedly, provided the solution to an old philosophical problem. An argument had been raging since the days of the Greek philosophers as to whether the events of this world are due to chance or to necessity. As far as evolution is concerned, Darwin put an end to this controversy. In short, owing to the two-step nature of natural selection, evolution is the result of both chance and necessity. There is indeed a great deal of randomness ("chance") in evolution, particularly in the production of genetic variation, but the second step of natural selection, whether selection or elimination, is an antichance process. The eye, for instance, is not a chance product, as so often claimed by anti-Darwinians, but the result of the favored survival of those individuals, generation after generation, who had the most efficient structures for vision. (For an expanded analysis, see Chapter 10.)

Another widespread erroneous view of natural selection must also be refuted: Selection is not teleological (goal-directed). Indeed, how could an elimination process be teleological? Selection does not have a long-term goal. It is a process repeated anew in every generation. The frequency of extinction of evolutionary lineages, as well as their frequent changes in direction, is inconsistent with the mistaken claim that selection is a teleological process. Also there is no known genetic mechanism that could produce goal-directed evolutionary processes. Orthogenesis and other proposed teleological processes have been thoroughly refuted (see Chapter 4).

To say it in other words, evolution is not deterministic. The evolutionary process consists of a large number of interactions. Different genotypes within a single population may respond differently to the same change of the environment. These changes, in turn, are unpredictable, particularly when caused by the arrival at a locality of a new predator or competitor. Survival during a mass extinction may be strongly affected by chance.

CAN NATURAL SELECTION BE PROVEN?

After one has fully understood natural selection as a population process, it seems so obvious that one is at once convinced of its correctness. This is indeed what happened to Charles Darwin. However, in 1859 when he published the *Origin*, he actually did not have a single clear-cut piece of evidence for the existence of selection. The situation has completely changed since then. In the nearly century and a half since 1859, great amounts of concrete evidence have been acquired (Endler 1986).

The response of the genotype to a selection pressure is sometimes extraordinarily precise, as in some cases of mimicry, but far less so in other situations. As Cain and Sheppard have shown, bandedness in the snail *Cepaea nemoralis* is in certain habitats advantageous over an unbanded shell, but it would be difficult to prove that five bands are selectively superior to three bands.

The first proof of selection was the discovery of mimicry. The tropical explorer Henry Walter Bates (1862) observed in Amazonia that some palatable species of butterflies had the same pattern and coloration as sympatric toxic or at least unpalatable species, and that wherever the noxious models varied geographically, the palatable mimics followed the same geographic variation (Fig. 6.1). This became known as *Batesian mimicry*. A few years later, Fritz Müller (1864) discovered that poisonous species also mimicked each other so that the insect-eating birds had to remember only one model to avoid, thus protecting three or four or even a dozen different toxic species. This greatly reduced predation loss in the toxic species that mimicked each other, since the young birds had to learn only a single pattern for a whole group of Müllerian mimics *(Müllerian mimicry)*.

Drug resistance of pathogens, as well as pesticide resistance in agricultural pests, eventually forced everybody to accept the importance of selection. In recent years, numerous occurrences of selection have been discovered by medical and public health workers. The relationship between the sickle cell gene and malaria resistance in Africa is a good example (Fig. 6.2 and Box 6.3). Industrial melanism, in which moths and other organisms adapt to polluted habitats by changing body coloration, is a phenomenon in which the occurrence of selection has been particularly well tested experimentally.

Box 6.3   Sickle Cell Gene and Human Hemoglobin

The human sickle cell gene demonstrates the drastic effects that a mutation may have, even if it results only in the replacement of a single amino acid. The sickle cell gene is common in most malarial regions, particularly in Africa, because it protects the heterozygous carrier against malaria. In the sickle cell mutation the amino acid glutamic acid in the (beta) globin chain is replaced by valine. The blood disease caused by this mutation is sooner or later fatal for the homozygous carrier, but the heterozygotes are protected against malarial infections. This advantage is lost when a carrier of the sickle cell gene moves into a malaria-free region, like the United States. The frequency of the sickle cell gene among the descendants of slaves is gradually being reduced owing to the mortality of the homozygous carriers without any recompensating advantage of being heterozygous.

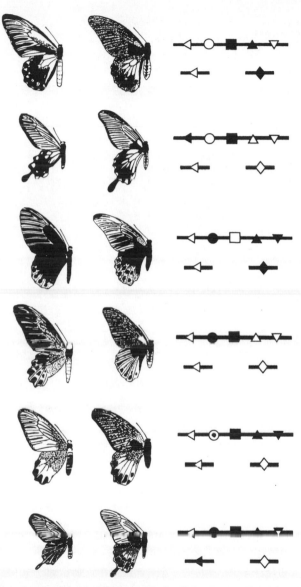

FIGURE 6.1

Geographic races of the Batesian mimic *Papilio memnon* (left) vary in parallel with the variation of their model (right). *Source*: Reprinted from the *Biology of Butterflies*, R. I. Vane and E. B. Ford, page 266, copyright © 1984, by permission of Academic Press, London.

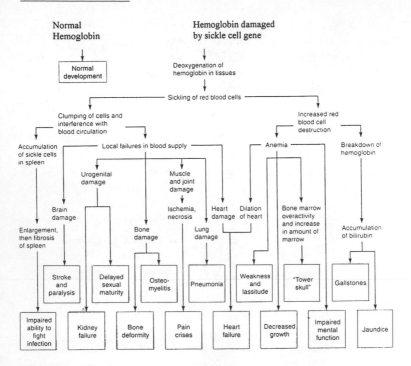

FIGURE 6.2

Pleiotropic effects of the sickle cell mutation. *Source*: Strickberger, Monroe M. (1985). *Genetics* 3rd ed. Prentice-Hall: Upper Saddle River, N.J.

STRUGGLE FOR EXISTENCE

Darwin chose the metaphor "struggle for existence" as the title of the third chapter of the *Origin*. Every individual, whether animal, plant, or other kind of organism, "fights" every minute of its life for survival. If it is potential prey, it struggles with predators; if it is a predator, it fights for prey against other predators. In order to survive an individual has to meet successfully all conditions of life. As Darwin said: "A plant on the edge of a desert is said to struggle for life against the drought, though more properly it should be said to be dependent on the moisture" (1859: 62). The plant that is superior in drought resistance to the other members of the same population will survive best. The competition is usually most severe among members of the same population; this competition is not only for food, but also shelter and all the needs for successful reproduction, such as territory and mates. And, as Darwin continues, "as more individuals are produced than can possibly survive, there must in every case be a struggle for existence" (1859: 63).

But such struggle takes place not only among members of the same species, but often among individuals of different species. For example, seed-collecting ants in the American West compete for plant seeds with rodents. Red squirrels compete with red crossbills for pine seeds. On pastures and salt marshes I have seen passing flocks of starlings compete with beautiful yellow-chested local meadowlarks. In the tidal zone there is a pitched struggle for space among barnacles, mussels, kelp, and other marine organisms. In many instances, two species with similar requirements may manage to coexist. Yet when one of the species is experimentally removed, the other may strikingly increase in number. Many other pairs of species cannot coexist because their requirements are too similar and one is a little superior. This is referred to as the *competitive exclusion principle*. Sometimes it is quite puzzling how two seemingly quite similar species can successfully coexist. In the Galapagos Islands, species of Darwin's finches coexisting on the same island have bills with different mean sizes and ranges of variation. If one of these species inhabits an island all alone, free of competition with the other species, its bill may have a much greater range of variation, including part of the range of variation of the species that competes with it elsewhere.

The importance of competition is demonstrated most graphically when a species becomes extinct as the result of an alien species successfully colonizing its range. Darwin called attention to the extinction of many native New Zealand species of animals and plants when introduced European species successfully established themselves there and outcompeted the natives.

Competition and other aspects of the struggle for existence exert an enormous selection pressure. Understanding the interactions among species has often been of great value for agriculture. Various pests of citrus orchards (aphids and scale insects) have been successfully controlled by ladybird beetles or other predatory insects. When introduced Opuntia cactuses were spreading like wildfire on Queensland sheep and cattle pastures, an Argentine moth (*Cactoblastis*) in no time virtually eliminated the cactus and restored tens of thousands of square miles of pastures to productivity. What these cases, and scores more in the ecological literature, demonstrate is that normally coexisting species live in steady-state harmony with each other, which is continuously adjusted by natural selection.

THE OBJECTS OF SELECTION

Who or what is being selected? Curiously this simple-sounding question has been the source of a long and continuing controversy. For Darwin, of course, as for virtually all naturalists since then, it was the individual organism that survived and reproduced. The genetics of the whole individual, however, cannot be dealt with mathematically, and most mathematical population geneticists, therefore, adopted the gene as the real "*unit of selection*." Other authors proposed still other putative targets of selection, such as groups of individuals or whole species.

Some students of animal behavior and some ecologists thought that selection acted to "improve" the species. Up to 1970 some geneticists still thought that not only genes but also populations were the units of selection. It was not until about 1980 that reasonable unanimity was reached that the individual is the principal target of selection.

Much confusion about this problem can be avoided by considering two separate aspects of the question: "selection of" and "selection for." Let us illustrate this with the sickle cell gene. For the question "selection of" the answer is an individual who either does or does not carry the sickle cell gene. In a malarial region the answer to "selection for" is the sickle cell gene, owing to the protection it gives to its heterozygous carriers. When one makes the distinction between the two questions, it becomes quite clear that a gene as such can never be the object of selection. It is only part of a genotype, whereas the phenotype of the individual as a whole (based on the genotype) is the actual object of selection (Mayr 1997). This does not reduce the importance of the gene in evolution, for the high fitness of a given phenotype may be due to one particular gene.

The reductionist thesis that the gene is the object of selection is also invalid for another reason. It is based on the assumption that each gene acts independently of all other genes when making its contribution to the properties of the phenotype. If this were true, the total contribution of genes to the making of the phenotype would be accounted for by the addition of the action of all individual genes. This assumption is referred to as the "additive gene action" assumption. Indeed some genes, perhaps even many genes, seem to act in such a direct and independent manner. If you are a male with the hemophiliac gene, you will be a bleeder. Many other genes, however, interact with each other. Gene B may enhance or reduce the effects of gene A. Or else the effects of gene A will not occur unless gene B is also present. Such interactions among genes are called *epistatic interactions*.

Obviously, epistatic interactions are not as easily determined as additive gene actions and their study has therefore been generally avoided by geneticists. One such interaction is designated as "incomplete penetrance." In such a case an individual may have a particular gene but does not show its effect, while it is fully reflected in the phenotype of another member of the population who has a somewhat different genotype. For instance, in a widely adopted model of the inheritance of schizophrenia, it is postulated that the major gene contributing to this illness has only a 25 percent penetrance, that is, is manifested in only 25 percent of the carriers of this gene. Some of the combinations of interacting genes are apparently tuned so finely that any deviation from the optimal balance is being selected against. Pleiotropy and polygeny are well-known instances of such interactions of genes (see Chapter 5).

The importance of this interaction of genes was not fully realized until the discovery of the regulatory genes, like hox and pax genes. With these genes we can observe very drastic interactions, but minor interactions among genes are very common. The question of what all of these interactions add up to is controversial. But there is a good deal of indirect evidence for the existence of an "internal balance" of the genotype or, as it has also been called, a "cohesion of the genotype." It has been postulated that this is a conservative element in evolution and accounts for the stasis in so many evolutionary lineages. It has also been postulated that this is the reason why founder populations may change drastically so often and so rapidly. Founder populations have a greatly reduced variance and may have a rather unbalanced set of genes. Such gene pools may respond to new selection pressures rather differently than the parent species and may be able to produce greatly diverging phenotypes.

It is important for the clarification of various evolutionary controversies to clearly understand how variable a contribution to fitness a gene may make. Many genes do not have a standard selective value. A gene may be beneficial when placed in one particular genotype, but it may be deleterious when placed in a genotype with different genes. Hence, the interaction of genes is of considerable importance for the selective value (fitness) of an individual. So-called neutral evolution (see below) is a meaningless concept considering the fact that the gene as such is not the target of selection.

A gene may have a very different effect on the fitness of an individual whether it is present in a single dose (heterozygote) or a double dose (homozygote). A single dose of the sickle cell gene greatly adds to the fitness of its heterozygous carrier in a malaria region, while in double doses (as in homozygotes) it is sooner or later lethal. This shows particularly graphically that a gene does not necessarily have a fixed selective value, but that this value may depend on the other genes with which it is associated in the genotype.

PHENOTYPE

What do we mean when we say that the individual is the object of selection? What is it that is encountered by natural selection and induces it either to favor or to disfavor an individual? Not its genes or genotype, because these are not visible to selection, but rather its phenotype. The word phenotype refers to the totality of morphological, physiological, biochemical, and behavioral characteristics of an individual by which it may differ from other individuals. The phenotype originates during development of the zygote from the fertilized egg to adulthood owing to the interaction of the genotype with its environment. The same genotype may produce quite different phenotypes in different environments. A semiaquatic plant, for instance, may produce entirely different leaves on land than in the water (Fig. 6.3)

FIGURE 6.3

Phenotypic variation in leaf form in the semiaquatic plant *Ranunculus aqu atilis*. Compare the filamentlike leaves on submersed branches (a) with normally structured leaves on branches above the water (b). *Source*: Herbert Mason, *Flora of the Marshes of California*. Copyright © 1957 Regents of the University of California, copyright renewed 1985 Herbert Mason.

The phenotype consists not only of the structure of an organism and of its physiology, but also of all the products of the behavioral genes. This includes the nest a bird builds, or the web of a spider, or the path of migration of a migratory bird. Dawkins (1982) has referred to these aspects of the characteristics of an organism as the extended phenotype. It is as much (and often more so) the target of selection as the structural characteristics of an organism.

The range of phenotypic variation that a particular genotype is able to produce is referred to as its *norm of reaction*. Thus the phenotype is the result of the interaction between genotype and environment. Some species have a very wide reaction norm; they can adjust their phenotype to wide variations of the environment and have a high phenotypic plasticity. The fact that it is the phenotype, rather than the genotype, that is the target of selection allows the existence of considerable genetic variation in a gene pool. Such variation is compatible with selection as long as the produced phenotypes have an acceptable selective value.

Because the phenotype is the product of the genotype, it has both evolutionary stability and evolvability. Many core processes at the cellular level are conserved throughout the metazoa, such as many signaling pathways and genetic regulatory circuits, others are conserved throughout the eukaryotes (e.g., the cytoskeleton), and yet others throughout all life-forms such as metabolism and replication. Sequence conservation is so strong that more than half of the coding sequences of yeast are recognizable in mice and humans. The actins of yeast and humans, for example, are 91 percent identical.

These core processes, however, must not be so structured that they prevent further evolution. Indeed there is ongoing selection for evolvability of the phenotype. Only this flexibility permits the occupation of new adaptive zones and the successful coping with new environmental challenges. The study of the properties of a genotype that allow it to cope with the constraints of the conserved portions and to maintain optimal evolvability is one of the frontiers of current evolutionary biology.

OTHER POTENTIAL OBJECTS OF SELECTION

The individual is not the only entity that has been suggested by one or another evolutionist to be the object of natural selection. We have already refuted this claim for the gene, and will now discuss gametes, groups, species, higher taxa, and clades.

Gamete Selection

All gametes are subject to selection between the completion of the meiotic cycle and either fertilization or death. Elimination is extremely severe since only a very small fraction of the gametes is successful. Unfortunately, we know very little about the eliminating factors. Experiments have revealed that the proteins of the egg wall in certain marine invertebrates have the capacity to prevent the entry of some spermatozoa while admitting others. What fitness criteria are involved here is still unknown. The properties responsible for gametic selection are important isolating mechanisms, referred to as gametic incompatibility.

Gamete interaction has been studied much more intensively in plants, particularly the compatibility reactions between pollen tubes and the stigma or style. In many taxa, special mechanisms prevent self-pollination. Less is known about incompatibilities among out-crossing species and how it is controlled. As early as the 1760s the botanist J. G. Kölreuter showed that when conspecific and alien pollen were placed on the stigma simultaneously, it was always the conspecific pollen that fertilized the seed. If the alien pollen alone was placed, it successfully fertilized in some species pairs.

Group Selection

There has been much argument whether a group of individuals can or cannot be the object of selection. The situation is clarified if one makes a distinction between "soft" and "hard" group selection (Mayr 1986). Soft group selection refers to selection of casual groups, and hard group selection to cohesive social groups. In the case of soft group selection, the fitness of the group is the arithmetic mean of the fitness values of the members of the group. This mean value has no effect whatsoever on the fitness of the composing individuals. The evolutionary success or failure of such a group ("*group selection*") is simply the automatic consequence of the fitness of the composing individuals. The fact that they are associated in a group makes no contribution to their fitness. Such soft group selection makes no independent contribution to evolution. This is the kind of "group selection" one finds in casual groups. Soft group selection really should not be referred to as group selection, because the group as such is not selected. A population as a whole is subject to such soft "group selection."

However, in certain species a special kind of group occurs, social groups, that can indeed be a target of selection. Such a group, owing to social cooperation among its members, has a greater fitness value than the arithmetic mean of the fitness values of its individual members. This may be called hard group selection. Members of such groups cooperate by warning of enemies, sharing newly discovered sources of food, and joint defense against enemies. This cooperative behavior enhances the survival propensity of such a group. The human species, at least during the hunter-gatherer stage, benefited from such social cooperation and this led to increased survival of certain groups. As a result, any genetic contribution toward cooperative behavior would be favored by natural selection. It is believed that this social cooperation has been an important factor in the development of human ethics (see Chapter 11). Hard group selection does not replace individual natural selection, but is superimposed on it.

Kin Selection

A form of selection called kin selection is recognized by many evolutionists, particularly in connection with the evolution of altruism. It is defined as selection for characteristics that favor the survival of close relatives of a given individual who shares part of the same genotype (known as inclusive fitness altruism). Except for parental care and in social insects, kin selection is probably not as important a factor in evolution as sometimes believed, particularly when there is a considerable exchange of individuals among neighboring groups. The altruism that members of a social group show to other related members of the group (excluding offspring) is apparently never anywhere near as great as the altruism displayed by parents (particularly mothers) to their own offspring. It is perhaps misleading to combine the two kinds of relationship under the single term kin selection. However, since members of a social group are often closely related to each other, much hard group selection is simultaneously kin selection. (See also Chapter 11.)

Species Selection

The history of evolution is a steady extinction of species and the origin of new species. This turnover is often apparently due to the superiority of a new species over an established species. Also, when species of different biota come into competition, as did those of North and South America after the isthmus of Panama had been established in the Pliocene, it can result in considerable extinction, part of it caused by competition between the invaders and the indigenous species. This phenomenon has been called species selection. As mentioned earlier, Darwin called attention to the frequent extinction of native species of plants and animals on New Zealand after European species had been introduced. The mistake was made by some authors to consider this an alternative to individual selection. In reality, this so-called species selection is superimposed on individual selection. The individuals of both species coexist after the entry in the same niche, and extinction takes place when the individuals of the invading species are, on average, superior to those of the indigenous species. Clearly, it is a selection of individuals. Misunderstandings are avoided if this process is called "species turnover" rather than "species selection" (see Chapter 10). A species as a unit is never the object of selection, only its individuals.

An even higher taxonomic level is involved in so-called clade selection, a clade being a holophyletic group of taxa forming a branch of a phylogenetic tree. Owing to the Alvarez extinction event at the end of the Cretaceous, the clade of the dinosaurs became extinct, but not the clades of birds and mammals. During every mass extinction certain higher taxa have fared better than others. Again, the actual objects of selection were individuals, but the individuals of some clades shared characteristics that favored survival through the extinction event, while these characteristics were absent in the individuals of the losing clades. What is remarkable for mass extinctions is that a whole higher taxon may be eliminated almost instantaneously or at least during a relatively short period. Clade extinctions also sometimes occur that are not clearly the result of a mass extinction. The extinction of the trilobites may be an example.

Competition Among Higher Taxa

Mass extinctions have called attention to the possible competition between higher taxa. Mammals had existed for some 100 million years prior to the mass extinction at the end of the Cretaceous, but they were small, insignificant, and quite likely nocturnal. Why did they enjoy such an explosive radiation in the ensuing period of the early Tertiary? The most widely accepted answer to this question is that they were able to enter all the ecological niches vacated by the demise of the previously dominant dinosaurs. Evidently the two classes of animals had been competing all along, but the dinosaurs had been superior competitors. Obviously, the mammals did not cause the extinction of the dinosaurs, but they replaced them when the dinosaurs became extinct owing to a nonbiological cause.

This case of the mammalian flourishing also illustrates the phenomenon of explosive speciation in previously vacant niches. Other examples are the species flocks of fish, molluscs, and crustaceans in ancient lakes and the rapid radiation of colonists of oceanic archipelagos. There are more than 700 species of drosophilid flies and over 200 species of crickets on the Hawaiian Islands. The honey creepers (Drepanididae) in the Hawaiian Islands and the geospizid finches in the Galapagos are other well-known cases of such radiation.

In all of these cases it was the absence or the removal of competition that made the radiation possible. One speaks of displacement when an incumbent taxon is exterminated by competitive exclusion owing to the arrival of a superior competitor. To actually prove the causal connection in such a sequence is difficult. For instance, the multituberculates were a flourishing group of nonplacental mammals in North America in the late Cretaceous and Paleocene. But when, in the Eocene, the first rodents appeared (probably from Asia) and became extremely successful, common, and widespread, the multituberculates became rare and finally died out. The extinction of the trilobites when the bivalves became so successful is another possible case, but an environmental catastrophe has also been proposed for their demise. Throughout the history of paleontology there are numerous similar cases of the sudden decline and ultimate extinction of a previously prosperous taxon after a new taxon with seemingly similar ecological requirements had appeared. It is, of course, impossible to prove in any of these cases that it was the arrival of the new competitor that caused the extinction, but this scenario often fits the known facts better than any other explanation.

WHY IS EVOLUTION USUALLY SO SLOW?

When the pharaohs' tombs were opened in Egypt early in the nineteenth century, not only human mummies were found but also those of sacred animals such as cats and ibises. When the anatomy of these animal mummies, estimated to be about 4000 years old, was carefully compared by zoologists with living representatives of these species, no visible differences could be found. This finding was in striking contrast to the rapidity with which animal breeders had produced pronounced changes in domestic animals in a much shorter time. Consequently, the absence of any visible change in these mummies was used as an argument against Lamarck's theory of evolution. Now we know that it usually takes many thousands if not millions of years for visible changes in evolving species to occur, except in a few special situations. Hence, the constancy of the Egyptian mummies is no evidence against evolution.

With drastic selection taking place in every generation, it is legitimate to ask why evolution is normally so slow. The major reason is that owing to the hundreds or thousands of generations that have undergone preceding selection, a natural population will be close to the optimal genotype. The selection to which such a population has been exposed is *normalizing or stabilizing selection*. This selection eliminates all of those individuals of a population who deviate from the optimal phenotype. Such culling drastically reduces the variance in every generation. And unless there has been a major change in the environment, the optimal phenotype is most likely that of the immediately preceding generations. All the mutations of which this genotype is capable and that could lead to an improvement of this standard phenotype have already been incorporated in previous generations. Other mutations are apt to lead to a deterioration and these will be eliminated by normalizing selection. There are also some special genetic mechanisms, such as *genetic homeostasis* (including heterozygote superiority), that favor maintenance of the steady state.

FOUNDER POPULATIONS
••••••••••••••••••••••••••••••

The genotype is a carefully balanced system owing to the epistatic interaction of the composing genes. Selection for the replacement of a gene by a new one may therefore require adjustments at other gene loci. The larger a population, the slower will be the incorporation and spread of new genes. By contrast, a small *founder population* established by the offspring of a single fertilized female or by a few founders may be able to shift more quickly to a new adaptive phenotype since it is unconstrained by the cohesive forces of a large gene pool.

There is much observational evidence that evolutionary change to the level of completed speciation proceeds more rapidly in peripheral populations than in large widespread species (Mayr and Diamond 2001). The explanation for this is still controversial. Dobzhansky and Pavlovsky (1957) (Fig. 6.4) showed a long time ago that a group of small, originally identical populations diverged from each other much more rapidly than a group of identical large populations. Other laboratory studies of founder populations did not find a drastic change in such populations. However, most of the studies were made with *Drosophila melanogaster*, a species with a phenotype that is apparently particularly stable as indicated by its various sibling species. This constancy of *D. melanogaster* raises the possibility that different founder populations may react differently to their isolation. Traditionally, the greater evolutionary inertia of large populations was ascribed to the greater amount of pleiotropy and polygeny. However, another cause is presumably the distribution of different regulatory genes. Conservative gene flow does not reach isolated populations and interfere with their increasing divergence. There is every reason to believe that new discoveries in developmental genetics will contribute a better understanding of the causes for different rates of evolutionary change in general and in speciation in particular.

WHAT IS THE EVOLUTIONARY ROLE OF BEHAVIOR?

For Lamarck, behavior was an important cause for evolutionary change. He thought that changes in organisms caused by any kind of activity would be transmitted to future generations by the inheritance of acquired characters. For example, when giraffes stretched their neck to reach higher leaves, the resulting elongation of the neck would be inherited by the next generation. Even though this theory of inheritance is now refuted, evolutionists still believe, but for very different reasons, that behavior is important in evolution. A change in behavior, for instance, adoption of a new food item or increased dis-

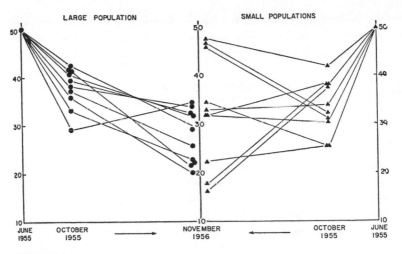

FIGURE 6.4

Variation, epistatic effects, and population size. The frequency (percent, vertical scales) of *PP* chromosomes in twenty replicate experimental populations of mixed geographic origin (Texas to California). The populations that have gone through a bottleneck of small population size show far greater variance after 17 months than the continuously large populations. *Source*: Mayr, F. *Animal Species and Evolution.* Belknap (HU Press), 1966.

persal, is apt to set up new selection pressures, and these may then lead to evolutionary changes (Mayr 1974). There are reasons to believe that behavioral shifts have been involved in most evolutionary innovations, hence the saying "behavior is the pacemaker of evolution." Any behavior that turns out to be of evolutionary significance is likely to be reinforced by the selection of genetic determinants for such behavior (known as the *Baldwin effect*).

SELECTION FOR REPRODUCTIVE SUCCESS (SEXUAL SELECTION)

When we speak of natural selection, unconsciously we always think of the struggle for existence. We think of the factors that favor survival, such as a capacity to overcome adverse weather conditions, to escape enemies, to better cope with parasites and pathogens, and to be successful in competition for food and habitation, in short to have any property that would enhance the chances for survival. This "survival selection" is what most people have in mind when they speak of natural selection.

Darwin, however, saw quite clearly that there was a second set of factors enhancing the probability of leaving offspring: all factors contributing to an increase of reproduction success. He called these factors *sexual selection*. Darwin listed here all cases of pronounced sexual dimorphism, such as the large antlers of deer, the magnificent tails of male peacocks, and the resplendent plumage of male birds of paradise and hummingbirds. Since females usually have the opportunity to choose their mates, those males will be favored by sexual selection who are most successful in attracting the favor of mate-seeking females. Other male characters that in some species are also favored by sexual selection are those that help males to be victorious in fights with rivals and that permit such males to acquire a larger harem of females, as occurs among seals, deer, sheep, and other species of mammals. Those males who benefit from such characteristics have increased reproductive success. However, enhanced reproductive success may also be effected by other means, such as the ability to obtain superior territories, sibling rivalry, aspects of parental investment, and other aspects of the interaction among individuals in a family or in a population. Darwin defined sexual selection as "the advantage which certain individuals have over others of the same sex and species solely in respect to reproduction." The term "selection for reproductive success" actually fits this broad definition better than the term sexual selection.

The cases in which competition or fighting among members of the same sex, such as among seal bulls or stags, leads to reproductive success have been designated "intrasexual selection." Other cases, such as female choice, in which the selection takes place between the two sexes, have been called "intersexual selection." There has been much discussion in recent years concerning the criteria employed by females when making their choice. Zahavi (1997) postulated that females may choose particularly conspicuous males, because their survival in spite of the handicap of conspicuousness indicates that they have superior qualities (the so-called *handicap principle*).

Simultaneous Mate and Niche Selection

One would expect particularly strong normalizing selection to achieve a complete constancy of the behavioral components of a species' isolating mechanisms to prevent loss of fitness through hybridization. This is indeed ordinarily true. However, it appears that in some cases mate selection is correlated with niche selection and that a heterogeneity of available niches may lead to a diversification of mate selection. Different kinds of males in a population may have different kinds of reproductive success in different subniches or habitats. In a species of freshwater fish, cichlids for example, some males may have a preference for benthic feeding and others for open-water feeding. Different kinds of females may originate in such a species, some with a preference for benthic and others for pelagic males. Eventually, through the process of sympatric speciation, two species may evolve. In such cases, sexual selection leads to speciation.

Sympatric speciation in plant-feeding insects may take place by simultaneous mate and host selections, as Guy Bush has maintained for many years. If insects that are on the whole host specific on plant A succeed in colonizing plant species B, and if the new colonists on B develop a mating preference for other individuals who have become adapted to species B, a host-specific new species will evolve on plant B, and reverse colonization from B back to A will be rare.

Sexual Dimorphism

Male and female appearances differ from each other in most animals, and there is an extraordinary range of variation in the degree of this sexual dimorphism. Certain deep-sea fish have dwarf males that are attached to the females, because free-swimming males might have difficulty finding females in these vast and rather lifeless spaces. At the other extreme, in certain species of seals, like the elephant seal, males may be several times larger than females because larger males can better defeat their rivals in territorial fights and so acquire larger harems. The magnificent plumages of male birds of paradise, hummingbirds, and other birds with polygamy were already mentioned under sexual selection. None of these cases poses any real difficulty for the theory of natural selection because all of these special male characteristics have the selective advantage of enhanced reproductive success. Invariably there is some counterselection against a runaway evolution of too extreme development of male characteristics. When they reduce survival, they will be selected against.

WHY DOES NATURAL SELECTION SO OFTEN FAIL TO ACHIEVE OR MAINTAIN ADAPTEDNESS?

Some enthusiasts have claimed that natural selection can do anything. This is not true. Even though "natural selection is daily and hourly scrutinizing, throughout the world, every variation even the slightest," as Darwin (1859: 84) has stated, it is nevertheless evident that there are definite limits to the effectiveness of selection. Nothing demonstrates this more convincingly than the fact that 99.99 or more percent of all evolutionary lines have become extinct. We must ask ourselves, therefore, why is natural selection so often unable to produce perfection? Recent studies have revealed numerous reasons for such limitation. A discussion of these constraints can contribute a great deal to the understanding of evolution. I recognize seven classes of such constraints.

1. *The limited potential of the genotype.* The existing genetic organization of an animal or plant sets severe limits to its further evolution. As Weismann expressed it, no bird can ever evolve into a mammal, nor a beetle into a butterfly. Amphibians have been unable to develop a lineage that is successful in salt water. We marvel at the fact that mammals have been able to develop flight (bats) and aquatic adaptation (whales and seals), but there are many other ecological niches that mammals have been unable to occupy. There are, for instance, severe limits on size, and no amount of selection has allowed mammals to become smaller than the pygmy shrew and the bumblebee bat, or allow flying birds to grow beyond a limiting weight.

2. *Absence of appropriate genetic variation.* A given species population can tolerate only a limited amount of variation. In the case of any drastic change of the environment, whether a climatic deterioration

or the appearance of a new predator or competitor, the kind of genes needed for an appropriate immediate response to this new selection pressure may not be present in the gene pool of the population. The high frequency of extinction documents the importance of this factor.

3. *Stochastic processes.* Much of the differential survival and reproduction in a population are not the result of selection, but rather of chance. Chance operates at every level in the process of reproduction, beginning with the crossing-over of parental chromosomes during meiosis to the survival of the newly formed zygotes. Furthermore, potentially favorable gene combinations are undoubtedly often eliminated by indiscriminate environmental forces such as floods, earthquakes, or volcanic eruptions before natural selection has had the opportunity to favor specific genotypes.

4. *Constraints of phyletic history.* Several alternate responses are usually possible to any environmental challenge, and it is the existing structure of the organism that often prescribes what response prevails. When the selective advantage of a skeleton developed among the ancestors of the vertebrates and of the arthropods, the arthropod ancestors had the prerequisites for developing an external skeleton, and the vertebrate ancestors for acquiring an internal skeleton. The entire evolution of these two large groups of organisms has since been affected by this choice among their remote ancestors. It permitted vertebrates to develop such huge creatures as dinosaurs, elephants, and whales, whereas a large crab is the largest type that the arthropods were able to achieve. The need for a regular molt of the external skeleton sets up in arthropods a formidable selection pressure against size increase.

Once a particular body structure has been acquired, it may not be possible to change it again. For instance, in terrestrial vertebrates the respiratory tract from the oral cavity to the trachea crosses the digestive tract, which also runs from the oral cavity to the esophagus. This arrangement was adopted in rhipidistian fishes, our aquatic ancestors. Although it poses forever the danger of food entering the trachea, no reconstruction of this inferior pathway occurred in several hundred million years.

Pelagic floating has been achieved by descendants of sessile, benthic, and actively swimming ancestors, belonging to many different animal phyla, who became adapted to the pelagic form of life through such very different adaptations as the inclusion of oil droplets, an increase of the surface area, and various other mechanisms. Each solution is a different compromise between the constraints or opportunities of this new adaptive zone and a species' previously existing physical structure. The adoption of a particular response to a new environmental opportunity may greatly restrict the possibilities of future evolution.

5. *A capacity for nongenetic modification*. The more pliable the phenotype is, that is, the larger its norm of reaction (owing to developmental flexibility), the more this reduces the force of an adverse selection pressure. Plants, and particularly microorganisms, have a far greater capacity for phenotypic modification than do higher animals. However, a capacity for nongenetic modification is present even in humans. This is exemplified by the physiological changes in a person when he or she ascends from the lowlands to high altitudes, where over the course of days and weeks the individual can become reasonably well adapted to the lowered atmospheric pressure and consequent decrease in oxygen. Natural selection is, of course, involved even in this phenomenon, since the capacity for nongenetic modification is under strict genetic control. Also, when a population shifts to a new specialized environment, genes will be selected during the ensuing generations that reinforce and eventually largely replace the capacity for nongenetic adaptation (the Baldwin effect).

6. *Unresponsiveness of the postreproductive age*. Selection cannot eliminate genetic propensities for diseases of old age. In the human species, for instance, most genotypes responsible for Parkinson's, Alzheimer's, and other afflictions that manifest themselves primarily in postreproductive life are relatively immune to selection. To some extent this is even true for diseases of middle age like prostate cancer and breast cancer, which usually strike toward the end of the active reproductive age.

7. *Developmental interaction*. It was realized by students of morphology as far back as Étienne Geoffroy St. Hilaire that there is competition among an individual's organs and structures. Geoffroy expressed this in his *La Loi de Balancement* (*Law of Balancing*, 1822). The different components of the morphotype are not independent of each other, and none of them responds to selection without interaction with the other components of the morphotype. The whole develop-

mental machinery is a single interacting system. An organism's structures and functions are compromises among competing demands. How far a particular structure or organ can respond to the forces of selection depends to a considerable extent on the resistance of other structures and other components of the genotype. Wilhelm Roux, more than 100 years ago, referred to the competitive developmental interactions as "the struggle of parts" in organisms.

The morphology of every organism reveals to what degree it is the result of a compromise. Every shift into a new adaptive zone leaves a residue of no longer needed morphological features that then become an impediment. One only needs to think of the many weaknesses in humans that are remnants of our quadrupedal and more vegetarian past, for instance, the facial sinuses, the structure of the lower vertebral column, and the caecal appendix. Such vestiges of former adaptedness are referred to as vestigial characters (see Chapter 2).

8. *The structure of the genotype.* The classic metaphor of the genotype was that of genes lined up like beads on a string. According to this view, each gene is more or less independent of the others and all of them are more or less similar in their nature. Not much is left of this view, which was generally accepted 50 years ago. To be sure, all genes are composed of DNA and the information they contain is coded in the linear sequence of base pairs. However, modern research in molecular genetics has revealed that there are different functional classes of genes, some charged to produce material, others to regulate it, and still others that are seemingly without function (see Chapter 5).

Furthermore, there is a good deal of rather indirect evidence that groups of genes may be organized into functional teams, which, in many respects, act as wholes (known as *modular variation*). However, this is a rather controversial area of molecular biology, and perhaps the best one can do at the present time is to call attention to the fact that the old "beads on a string" image of the genotype is no longer valid, and that there is still great uncertainty as to the actions of the genotype. The fact that there are transposons, introns, middle repetitive DNA, highly repetitive DNA, and many other kinds of noncoding DNA suggests different functions, but most of what these elements are and how they work together is still to be determined. More will be learned about the process of evolution as our understanding of the structure and functioning of the genotype improves than by anything else.

THE ROLE OF DEVELOPMENT IN EVOLUTION
....

The fertilized egg, the zygote, is a formless mass. It is converted into the phenotype of the adult stage during the embryonic or larval stages of development. Changes occurring during development are responsible for the divergence of different evolutionary lineages. Hence the study of development, of the ontogeny of the developing zygote, is of major concern to every evolutionist. However, the methods of classic embryology and in particular those of experimental embryology (*Entwicklungsmechanik*) were not suitable to produce the needed synthesis between embryology and genetics; this is finally being achieved by molecular biology. What was needed was a study of gene action, that is, a determination of the contribution to the development of the embryo made by each gene. This led to the discovery of the great diversity of genes, and in particular to the discovery of the regulatory genes (see Chapter 5).

Development is rarely direct. In a high proportion of animals the adult stage is reached through one or several larval stages, some of them requiring highly specific adaptations. One needs only to think of the caterpillar and butterfly or of the planktonic larva of the barnacle and its mollusclike adult. In these cases, new adaptations are acquired by some ontogenetic stages, but in other cases, particularly among parasites, certain phenotypic adaptations of the adult stage are lost, as in the Sacculina parasite of certain crabs.

DEVELOPMENT
....

Evolutionists, stretching all the way back to Darwin, realized that the "type" does not evolve as a unit and at the same rate in all of its parts, but that some components of the phenotype evolve faster, and some more slowly. This can be observed when a phyletic lineage shifts from one adaptive zone to a different one. Archaeopteryx, the earliest well-known fossil bird, had already acquired various avian characters—feathers, wings, the capacity for flight, enlarged eyes, and a birdlike brain—but had retained a reptilian stage of other parts of its structure (teeth, tail vertebrae). Such an unequal rate was referred to in earlier chapters as mosaic evolution. It would seem in such cases as if the phenotype was produced by more or less independent sets of genes. Consequently it has been postulated that the genotype is composed of an assemblage of gene modules, each controlling one of the mosaics of the phenotype. This thought was quite unpalatable to highly reductionistic geneticists, but the evidence for a somewhat modular structure of the genotype is increasing. If this is correct, a single regulatory gene may control such a module of genes. In other words, a mutation of the regulatory gene may result in a rather drastic change (discontinuity) of the phenotype. In other cases such a module may consist simply of a set of genes temporally brought together by selection for a particular state of adaptedness, but that might again be disassembled when the selective conditions change. There is a lot of structure in a genotype that cannot be discovered and explained by a purely reductionistic approach.

A Balance of Selection Pressures

No individual is ever perfectly adapted, as was stressed early on by Darwin. The main reason for this is perhaps that every genotype represents a compromise of genetic variability and stability. Most environments are perpetually changing, and at the end of a drought period a population will be better adapted for drought conditions than for an oncoming wet period. In the long run, the genotype strikes a balance between conflicting demands. And the same is true for the behavior of an organism toward predators and competitors. Mathematically inclined evolutionists have expressed this in terms of the game theory and superior strategies. Actually, of course, animals hardly test the various strategies in their minds. Rather their genotype predisposes some individuals in a variable population to be more timid and others to be more bold. Those individuals with the most successful balance of the two tendencies in a given situation will have the best chance to survive. There is no selection for the favorite type, rather the mean value of the population will reflect the balance of success of the various, sometimes rather conflicting genetic tendencies.

The response to a change in environmental conditions is often not predictable. When the climate of North America became more arid in the Pliocene, the vegetation responded and grasses, indeed eventually rather harsh, unpalatable grasses, took over. The browsing species of horses became extinct and were replaced by hypsodont species (see Chapter 10). When later a mesic period returned, several species of horses shifted back to browsing but retained their high teeth. In other cases, the return to an earlier environmental condition will be answered by a reversal of the selection. When industrial pollution was drastically reduced in recent years, the frequency of the black phenotype of the peppered moth (*Biston betularia*) also drastically declined in parallel with the reduction of soot and sulfur dioxide.

....

CHAPTER 7

ADAPTEDNESS AND NATURAL SELECTION:
ANAGENESIS

How can we explain why organisms are so remarkably well adapted to the environment in which they live? When preoccupied with other thoughts, we take this adaptedness very much for granted. Of course, a bird has wings to fly with and other attributes that are needed for its aerial existence. Of course, a fish has a streamlined shape and fins to enable it to swim; it has gills to take up the needed oxygen. So it is with all the properties of adapted organisms. But when you think about this more deeply, you begin to wonder how this admirable world of life could have reached such astonishing perfection. By perfection I mean the seeming adaptedness of each structure, activity, and behavior of every organism to its inanimate and living environment. Examples of such seeming perfection are structures such as the eye of vertebrates and insects, or the annual migrations of birds to their tropical winter quarters and their return with extraordinary precision to the spot from which they had started the previous autumn, or the admirable cooperation of the members in a colony of social insects, such as ants or bees.

As far back as we have written records, an occasional thinker or founder of a religion would ask these questions of why and how. Before the rise of science, only revealed religion could give an answer. Indeed, during the seventeenth and eighteenth centuries, adaptations were considered by the pious to be proof for the existence of a wise creator who had designed every created organism with the appropriate structures and behaviors needed for its particular place in nature (e.g., William Paley). Natural theology, the study of the work of the creator, was considered a branch of theology. This interpretation of the design of the living world is still defended in this age of science by the creationists.

Yet the claims of natural theology ran into considerable difficulties. Yes, wolves kill sheep, but it was argued that the creator had created sheep specifically so that the wolves would not die of starvation. A closer study of living nature, however, revealed an alarming amount of brutality and waste. As scientists came to understand more and more about the natural world, the credibility of perfect design by a benign creator further declined. Consideration of how God could have carried out his task of Creation raised even more serious difficulties. The manifold adaptations of structure, activity, behavior, and life cycle for each of the millions of species of organisms were far too specific to be explained by general laws. On the other hand, it seemed quite unworthy of the creator to believe that he personally arranged every detail in the traits and life cycles of every individual down to the lowest organism. The analysis of parasitism and other seemingly rather cruel aspects of the living world added to the loss of credibility. It came as a considerable relief for the thinking naturalists of the nineteenth century when they were able to replace the supernatural explanation of natural theology by a naturalistic explanation. However, to find a workable naturalistic explanation turned out to be a very difficult task.

The process of adaptation fitted very well with the thinking of natural theology and with Aristotle's belief in a "final cause." Adaptation in the non-Darwinian orthogenetic theories of evolution was attributed to intrinsic final causes. Even after 1859 many antiselectionist evolutionists still considered adaptation to be a more or less finalistic process. Actually there is no trace of any finalistic factor in the Darwinian explanation of the process of adaptation.

Darwin proposed an explanation of *adaptation*, based on population thinking, that succeeded in refuting all attacks since made against it. This was the application of the theory of natural selection to the process of adaptation (see Chapter 6), in which a character of an organism is an adaptation when among the variable populations of the ancestors it had been favored by nonelimination. The process of eliminating the less well adapted organisms results in the survival of the better-adapted individuals. Since this is equally true for the offspring of every set of parents in the population, the population as a whole remains well adapted or, perhaps, even increases in adaptedness.

DEFINITION OF ADAPTATION
••••••••••••••••••••••••••••••••••

There must be literally hundreds of definitions of adaptation in the literature. Ultimately, most agree that a trait is adaptive if it enhances the fitness (however defined) of an organism, that is, if the trait contributes to the survival and/or better reproductive success of an individual or social group. Or: an adaptation is a property of an organism, whether a structure, a physiological trait, a behavior, or any other attribute, the possession of which favors the individual in the struggle for existence. We believe that most such traits were acquired by natural selection or, if they arose by chance, their maintenance was favored by selection.

In determining what qualifies as an adaptation, it is the here and now that counts. It is irrelevant for the classification of a trait as an adaptation whether it had the adaptive quality from the very beginning, like the external skeleton of the arthropod, or acquired it by a change of function, like the swimming paddle of a dolphin or a *Daphnia*. One must always remember that adaptation is not a teleological process, but the a posteriori result of an elimination (or of sexual selection). Being an a posteriori process, the earlier history of a piece of the phenotype is therefore of little relevance for its adaptive value. The recognition of an adaptation is facilitated if it also occurs in other preferably unrelated organisms living in a similar environment, or if the adaptive quality of the character can be modified by appropriate experiments. One way to assess adaptations is by studying the variation of the adaptive character in variable natural populations. For an analysis of the problem of how to define adaptation, see West-Eberhard (1992) and Brandon (1998).

WHAT IS THE MEANING OF THE TERM ADAPTATION?

Unfortunately, the word adaptation is used in the evolutionary litera-ture for two entirely different subjects, one legitimate and the other not. This has created a great deal of confusion.

The legitimate use of the term adaptation is for a property of an or-ganism, whether a structure, a physiological trait, a behavior, or any-thing else that the organism possesses, that is favored by selection over alternate traits. But the term also has been used quite incorrectly for the process ("adaptation") by which the favored trait was actively acquired. This view can be traced back to the ancient belief that or-ganisms had an innate capacity for improvement, for steadily becom-ing "more perfect." Also, if one accepts an inheritance of acquired characters, activities such as the straining of the neck by giraffes "adapts" the neck to an improved construction. In this view, adapta-tion is an active process with a teleological basis. Some recent authors still seem to look at adaptation as such a process and therefore reject the whole concept of adaptation. But this is not defensible.

Adaptation is a completely a posteriori phenomenon for a Dar-winian, that is, it is based on the inductive assessment of facts. In ev-ery generation, all individuals that survive the process of elimination are de facto "adapted" and so are their properties that enabled them to survive. Elimination does not have the "purpose" or the "teleologi-cal goal" of producing adaptation; rather, adaptation is a by-product of the process of elimination.

To avoid the ambiguity of the word adaptation, it is preferable to use the word adaptedness for the state of being adapted. There is, however, no reason not to use the term adaptation for a property ac-quired or maintained by natural selection because it provided supe-rior survival chances in competition with other individuals. Many adaptations acquire a new role through change of function such as the swim bladder of fishes from lungs, or the middle ear bones of mam-mals from bones of the reptilian jaw articulation. The process of adaptation is a strictly passive one. Individuals that do not have as good an adaptation as others are eliminated, but the survivors do not contribute to the process of becoming better adapted by any special activities, as proposed in teleological theories of evolution. It is not

Box 7.1 Low Fertility of the Large Albatrosses *(Diomedea)*

Characteristic	Albatross	Most birds
Number of eggs	1	2-10 plus
Age at first reproduction	7-9 years	1 year or less
Sexual cycle	2 years or more	1 year or less
Life expectancy	Estimated to be 60 years or more	Mostly less than 2 years

particularly helpful to make a terminological distinction between adaptations that previously had a different role and those that originated as a consequence of the role they still fulfill. In addition to having specific adaptations, an organism as a whole is adapted to its environment.

The adaptations for optimal reproductive success possessed by certain species are quite astonishing. The large albatrosses of the waters of the Southern Ocean have only a single young every second year and do not enter the age of reproduction until they are seven to nine years old. How could natural selection have led to such a reduction of fertility? It was found that only the most able and experienced birds can find enough food to raise their young in this zone of incessant, powerful storms. On the other hand, they have the advantage of being able to establish breeding colonies on predator-free islands without any serious competitors. Hence the delay in age of reproduction and a reduction in the number of offspring were of selective advantage. The breeding cycle of the Emperor Penguin is another example. These birds court and lay their single egg under the most adverse conditions at the beginning or in the middle of the Antarctic winter, a season of frequent blizzards. The advantage of this timing is that the young hatch at the beginning of the southern spring and are raised during the southern summer, when conditions for their survival and growth are at an optimum. Such drastic reduction of fertility in albatrosses and penguins is compensated by an increased longevity of the adults and by the absence of predators from their breeding colonies on islands or on the Antarctic ice. Adaptations of extreme specialists, such as parasites, are sometimes even more astonishing.

TO WHAT IS AN ORGANISM ADAPTED? WHAT IS A NICHE?

We commonly say that a species is adapted to its environment. But this is not a sufficiently precise answer. A species shares its environment with hundreds of other species. For a hummingbird of the tropical forest who feeds in the canopy and builds there her nest, it is irrelevant whether or not some rocks are lying on the forest floor. Every species is adapted to a rather restricted selection of properties of the environment. These properties are certain general conditions (mostly climatic), but also specific resources (food, shelter, etc.). This specific set of environmental properties provides a species with the required living conditions called its *niche*. There are two ways to define a niche. The classic way is to consider nature to consist of thousands and millions of potential niches occupied by the various species adapted to them. In this interpretation, the niche is a property of the environment. Some ecologists, however, consider the niche to be a property of the species that occupies it. For them the niche is the outward projection of the needs of a species.

Is there a way to determine which of the two concepts of the niche has greater validity? The following example may help us to make up our mind. The large Sunda Islands, Borneo and Sumatra, west of Wallace's Line, each have about 28 species of woodpeckers. Even though the tropical rain forest of New Guinea, east of Wallace's Line, is remarkably similar to that of the Sunda Islands, with many of the dominant trees even belonging to the same genera, there is not a single woodpecker in New Guinea. Does this mean that there is no woodpecker niche in New Guinea? Definitely not! If we make a detailed analysis of the niches of the Malayan woodpeckers, we find that many of them are matched by analogous constellations of environmental factors in New Guinea. It would be quite misleading therefore to say that there are no woodpecker niches in New Guinea. Actually, the open niches are virtually calling for them, but woodpeckers are notoriously poor in crossing water gaps, and they simply did not succeed in crossing the various large water gaps between Sulawesi and New Guinea. And none of the indigenous families of New Guinea birds initiated a "woodpecker" branch. Many other pieces of evidence show that the classic definition of the niche, as a property of the environment, is preferable to the one that considers it a property of the organism. Biogeographers know that every colonizing species has to become adapted to the prospective niches it encounters in a newly occupied area. The word environment itself is often used in two very different senses, for all the surroundings of a species or biota or only for the niche-specific components.

LEVELS OF ADAPTATION

It is useful to distinguish between different levels of adaptation—adaptation for broad adaptive zones and adaptation for species-specific niches. Adaptations are hierarchically organized at different levels. This makes a specialization for highly specific niches possible. Among birds, we recognize woodpeckers, tree creepers, raptors (diurnal and nocturnal), waders (of a great range of sizes), swimmers, divers, terrestrial runners (ostrich, roadrunner), fish eaters, carrion eaters, seed eaters, and nectar feeders. They all have special adaptations of their bills, tongues, legs, claws, sense organs, digestive organs, and other structures and behaviors. These are mostly related to their mode of feeding or locomotion. All of these are adaptations for the special niches that these different kinds of birds occupy. And all are compatible with the demands of the special adaptive zone that birds occupy, namely, the air space. They differ from reptiles, their ancestors, by numerous adaptations for flying. They have feathers and wings, have reduced their weight by the loss of teeth and the tail vertebrae, and have hollow, thin-walled bones. They are endothermic and have numerous physiological adaptations for flight.

General and Special Adaptations

When we study the lifestyle of any particular group of organisms, we are at once impressed by the presence of very specific adaptations that make this lifestyle possible. Every book on animals describes such adaptations. Birds, for instance, have wings, feathers, lost the heavy teeth, have hollow bones, lost the bony tail, are endothermal, and possess physiological adaptations for flight. However, as Darwin already emphasized, birds have a second set of characters, all of which they share with other vertebrates, and which they inherited from their ancestors. These are not special adaptations for flight but are aspects of their vertebrate body plan. The genes for this part of the avian phenotype are components of the basic developmental machinery of birds derived from their ancestors, and in its totality it is adaptive, but not reducible into separate characters.

During embryonic development the basic features of the body plan are laid down before the special adaptations for their niches begin to develop. This explains all cases of so-called recapitulation (remember the age-old mantra "ontogeny recapitulates phylogeny"), such as the development of teeth in whale embryos or of gill arches in terrestrial vertebrates. An organism has to be well adapted as a whole, but it also must be able at all times to cope with its ancestral genome. Not every part of an organism is an ad hoc adaptation for its present lifestyle. These ad hoc adaptations are superimposed on the basic body plan. Nothing illuminates this better than the fact that in the ocean one can find representatives of as many as 15 or 20 phyla happily coexisting in the same general area. The enormous differences in their body plans do not prevent their perfect adaptation to their environment.

THE ADAPTATIONIST PROGRAM: CAN ONE PROVE ADAPTEDNESS?

How can one prove that certain individuals, as well as their structures and behaviors, are truly well adapted? This is a valid and indeed a very important question. It can be answered mainly by the ever repeated and severe testing of the supposedly adaptive attributes of organisms. This is the so-called adaptationist program outlined below (Gould and Lewontin 1979). For a refutation of the Gould and Lewontin critique of the adaptationist program, see Mayr (1983), Brandon (1995), and West-Eberhard (1992).

In an adaptational analysis it is of particular importance to consider the numerous constraints (Mayr 1983) that usually prevent a component of the phenotype from reaching optimal adaptedness. It must always be remembered that the individual as a whole is the target of selection and that there is an interaction between the selection pressures on different aspects of the phenotype. This is well illustrated by *Archaeopteryx*, which first acquired the most immediately needed flight adaptations—feathers, wings, improved eyes, enlarged brain— but was still not yet fully flight adapted in the retention of some less important reptilian characters (teeth, tail).

There are theoretically two ways to supply proof for the adaptedness of a feature. First, one can try to show that the occurrence of the feature cannot possibly be explained by chance. But it is very difficult to succeed in this endeavor. Second, one can test the various possible adaptive advantages of the feature, and its adaptedness is confirmed when all attempts to disprove these advantages are unsuccessful. What must be tested is the adaptedness of the particular phenotypic feature in question.

Almost any feature of an organism can be and has been shown to be of selective significance. Cases that have been experimentally tested are industrial melanism, banding patterns in snails, mimicry, aspects of sexual dimorphism, and scores of others reported in the literature (Endler 1986). By contrast, it is virtually impossible to prove that any property of an organism is *not* of selective significance. One is therefore forced to apply the second method and adopt the chance explanation only when all endeavors to demonstrate a selective value of a feature have been failures.

Adaptedness Is Acquired Gradually

New adaptations are ordinarily acquired quite gradually. *Archaeopteryx*, a 145-million-year-old fossil bird, documents almost perfectly the intermediacy between reptiles and birds. It still had teeth, a long tail, simple ribs, and the separated ilia and ischia of a reptile, but also had the feathers, the wings, the eyes, and the brain of a bird. The fossil ancestors of whales document a similar intermediate state in their adaptation to two different media. Darwin marveled that such a wonderful structure as an eye could have evolved through natural selection, but the comparative anatomists have shown not only that eyes evolved in the animal series at least 40 times independently, but also that among the existing photosensitive organs every intermediate step is found between a simple light-sensitive spot on the epidermis and a perfect eye with all of its accessories. The same regulatory gene *(Pax 6)* occurs in all forms with eyes, but is also widespread in eyeless taxa. It is apparently a very old regulatory gene that has been coopted for vision whenever eyes were selected.

Convergence

Open ecological niches or zones are often repeatedly colonized by entirely unrelated organisms that, once adapted to these niches, become by convergence extremely similar. The outstanding example is the Australian fauna of marsupial mammals, which, in the absence of placental mammals, have evolved adaptive types corresponding to (and remarkably similar to) Northern Hemisphere placentals such as the flying squirrel, mole, mouse, wolf, badger, and anteater. Very similar but unrelated nectar-feeding birds have evolved in Australia (honeyeaters), Africa and India (sunbirds), Hawaii (honeycreepers), and the Americas (hummingbirds) (see Fig. 10.4); ratites, the flightless birds with rudimentary wings, in South America, Africa, Madagascar, Australia, and New Zealand; and tree creepers in Australia, the Philippines, Africa, the Holarctic, and South America. The unrelated American and African porcupines are so similar that until recently they were considered to be closely related. Similar cases of convergence can be found in almost all groups of animals and even in plants (e.g., American cactuses and African euphorbs, see Fig. 10.5). Even only distantly related animals, like sharks (fishes), ichthyosaurs (reptiles), and porpoises (mammals), have become superficially very similar to each other.

The ubiquity of adaptedness is also documented by plants, fungi, protists, and bacteria. Life-forms have an astonishing capacity to vary, to respond to natural selection, and to take advantage of ecological opportunities.

CONCLUSIONS
....

Evolution in sexually reproducing organisms consists of genetic changes from generation to generation in populations, from the smallest local deme to the aggregate of interbreeding populations in a biological species. Numerous processes, particularly mutation, contribute to these genetic changes to supply the phenotypic variation needed by selection. The most important factor is recombination, which is largely responsible for the virtually inexhaustible supply of new genotypes in every generation. Selection, then, is responsible for the elimination of all but on the average two of the offspring of two parents. Those individuals that are best adapted to the abiotic and biotic environment have the greatest chance to be among the survivors. This process favors the development of new adaptations and the acquisition of evolutionary novelties, thus leading to evolutionary advance, as stated in the language of evolutionary biology. Evolution, being on the whole a population turnover, is ordinarily a gradual process, except for certain chromosomal processes that may lead to the production of a new species-individual in a single step.

Genetic material (nucleic acids) is constant and impervious to any influence from the environment. No genetic information can be transmitted from proteins to nucleic acids, and so the inheritance of acquired characters is therefore impossible. This provides an absolute refutation of all Lamarckian theories of evolution. The Darwinian model of evolution, based on random variation and natural selection, explains satisfactorily all phenomena of evolutionary change at the species level, and in particular all adaptation.

III
ORIGIN AND EVOLUTION
OF DIVERSITY:
CLADOGENESIS

··

CHAPTER 8

THE UNITS OF DIVERSITY: SPECIES

The early naturalists of Europe had no idea of the overwhelming richness of the world's organic diversity. The more conspicuous animals and plants of their neighborhood were all they knew. But this changed rapidly after the Middle Ages. The exploring voyages of the sixteenth to nineteenth centuries revealed the fact that each continent had an indigenous biota and also that there were great latitudinal differences, with the tropics having a very different life from the temperate and arctic regions. Oceanic research revealed a rich marine life, from the surface down to the greatest ocean depths, and the microscope disclosed the enormous world of planktonic and soil eukaryotes, small arthropods, algae, fungi, and bacteria. And this was not the end of the discoveries. Paleontology added an entire new dimension, the life of past geological periods.

This is not the place for a review of the enormous achievement of taxonomy to have described and classified nearly four million species of organisms (with somewhere between five and twenty million species still remaining undescribed). Instead I will focus on an explanation of the evolutionary aspects of this amazing diversity.

HOW MANY SPECIES OF LIVING ORGANISMS?
···

Few nonspecialists realize how difficult it is to answer this question. First of all, the agamospecies of asexual organisms, particularly of Prokaryotes, are something entirely different from the biological

TABLE 8.1   Number of Described Living Species (in thousands)

Kingdoms		Selected Phyla or Classes	
Protozoa	100	Vertebrates	50
Algae	300	Nematodes	500
Plants	320	Molluscs	120
Fungi	500	Arthropods	4,650
Animals	5,570	(crustaceans	150)
	6,790	(arachnids	500)
		(insects	4,000)

SOURCE: From May (1990).

species of the sexually reproducing taxa. More importantly, the majority of taxa are still poorly known. It occurs commonly that in a revision of a tropical genus of insects or spiders, 80 percent of the recognized species are new to science. The same is true for nematodes, mites, and numerous obscure groups. In 1758 Linnaeus knew some 9,000 species of plants and animals. By now about 1.8 million species of animals have been described (excluding agamospecies) and the grand total of species is estimated to be at least 5 to 10 million. Most of these live in the canopy of the tropical rain forest and, with 1–2 percent of this forest being destroyed every year, this number will soon be reduced appreciably.

The figures suggested by Robert May in Table 8.1 are very conservative. They are based on the biological species concept. If one uses instead a typological (including phylogenetic) species concept (see below), one can more than double these figures. May's figures are also low because they do not allow for sibling species. A figure of 5.57 million species for living animals is surely too low, but other estimates that range as high as 30 million are surely much too high. The greatest value of these figures is for comparative purposes. For example, the terrestrial warm-blooded mammals have less than half as many species (4,800) as the warm-blooded aerial birds (9,800 species) (Table 8.2).

Mammals and birds are best known, yet even in birds about three new species are discovered each year, and in mammals, in addition to bats and rodents, spectacular new large mammals were recently discovered in Vietnam. The figure of 9,800 species for birds is based on a liberal interpretation of polytypic species, in which peripherally iso-

TABLE 8.2   Number of species in Major Classes of Vertebrates

Teleost fishes	27,000
Amphibians	4,000
Reptilians	7,150
Birds	9,800
Mammals	4,800

lated populations are mostly listed as subspecies (see Fig. 8.1 for an example). If many of these were ranked as allospecies, the number of bird species could rise to 12,000. By far the largest group of animals are the beetles. For many families of animals, even for some orders and classes, there is at the present time not a single specialist in the world. It is feared that the description of the hitherto unknown species of organisms will proceed in the future at a slower rate than in the past. For a survey of this problem see May (1990).

Naturalists have long been faced by a puzzling conflict. On one hand, there is a pervasive continuity in the gradual change of the populations of a species through time and space and, on the other hand, there are gaps between all species and all higher taxa. Nothing has more impressed the paleontologists than the discontinuous nature of the fossil record. This is the reason why so many of them were such strong supporters of saltational theories of evolution. However, because we now know that saltations do not occur, we must ask the question: How do the gaps between species originate?

SPECIES CONCEPTS AND SPECIES TAXA
....

Obviously one cannot study the origin of gaps between species unless one understands what species are. But naturalists have had a terrible time trying to reach a consensus on this point. In their writings this is referred to as "the species problem." Even at present there is not yet unanimity on the definition of the species. There are various reasons for these disagreements, but two are most important. The first is that the term species is applied to two very different things, to the species

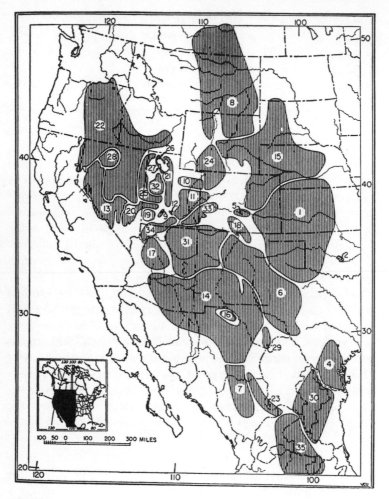

FIGURE 8.1
Polytypic species. The distribution of 35 subspecies of the kangaroo rat *Dipodomys or-
dii* Woodhouse. Numbers designate the ranges of various subspecies. *Source:* Mayr
1969.

as concept and to the species as taxon. A species concept refers to the
meaning of species in nature and to their role in the household of nature. A species taxon refers to a zoological object, to an aggregate of
populations that, together, satisfy the definition of a species concept.

The taxon *Homo sapiens* is an aggregate of geographically distributed populations that, as a whole, qualify under a particular species concept (see below). The second reason for "the species problem" is that within the last 100 years most naturalists have changed from an adherence to the typological species concept to acceptance of the biological species concept.

If the differences among the populations throughout the geographic range of a species are minor, not justifying taxonomic recognition, a species is called monotypic. Quite often, however, certain geographic races of a species are sufficiently different to be recognized as subspecies. A species taxon consisting of several subspecies is called a polytypic species (Fig. 8.1).

Species Concepts

Traditionally any class of objects in nature, living or inanimate, was called a species if it was considered to be sufficiently different from any other similar class. Such a species has a number of species-specific characteristics by which it can be distinguished from other species. Philosophers referred to such species as "natural kinds." This species concept, in which the species is considered to be a well-circumscribed class, is called the *typological species concept*. According to this concept, a species is a constant type, separated from any other species by an unbridgeable gap. In sexually reproducing species at a given time, it is usually not difficult to sort the organisms one finds at a given place into different species. One refers to the stated conditions as the "nondimensional situation." Such species coexist at the same time and at the same place and are usually separated from each other by a well-defined discontinuity.

Toward the end of the nineteenth and at the beginning of the twentieth century, more and more naturalists realized that species of organisms are not types or classes, but rather are populations or groups of populations (see Chapter 5). Also, it was found that the basic operational principle of the typological species concept—"species status is determined by the degree of phenotypic difference"—ran into practical difficulties. For instance, *sympatric* natural populations were found ever more frequently that did not interbreed even though they showed no evident taxonomic differences. This did not fit the typological species definition at all. Such species are now called *cryptic* or *sibling species*. These species show the same genetic, behavioral, and ecological differences from traditional species as do phenotypically different species but do not possess the traditional taxonomic differences. Sibling species also occur in plants (Grant 1981) and protists.

SIBLING SPECIES

Coexisting species that do not differ by noticeable taxonomic characters are remarkably common. Malaria in Europe had a very puzzling pattern of distribution until it was discovered that the malaria mosquito, *Anopheles maculipennis*, was actually an aggregate of six different sibling species, some of them not vectors of the malaria parasite. The famous protozoologist T. M. Sonneborn worked for over 40 years on the ciliate *Paramecium aurelia* and its varieties, until he realized that it consisted of 14 sibling species. Almost 50 percent of the North American species of crickets were discovered only by their different songs, they are that similar to each other. So far, very little is known about the frequency of sibling species in most phyla and classes of animals (see Box 8.1).

The discovery of what one might consider the opposite situation was equally disturbing for the typological taxonomist. One found in many species individuals that were strikingly different from other members of their population, yet they reproduced successfully with them. The Blue Goose and Snow Goose complex is one example; scores of others are cited in Mayr (1963: 150–158). Both of these situations did not at all fit the typological species definition.

Taxonomists finally came to the conclusion that they had to develop a new species concept, not based on degree of difference but on some other criterion. Their new concept was based on two observations: (1) species are composed of populations, and (2) populations are conspecific if they successfully interbreed with each other. This reasoning resulted in the so-called *biological species concept (BSC):* *"Species are groups of interbreeding natural populations that are reproductively isolated from other such groups."* In other words, a species is a reproductive community. Its reproductive isolation is effected by so-

---

Box 8.1   Sibling species

Sibling species are natural populations that are reproductively isolated from each other even though they often coexist sympatrically without interbreeding. Yet they are totally or virtually indistinguishable by traditional taxonomic characters. They are remarkably common in many higher taxa.

---

called *isolating mechanisms*, that is, by properties of individuals that prevent (or make unsuccessful) the interbreeding with individuals of other species.

ARE THERE OTHER SPECIES CONCEPTS AND DEFINITIONS?

In the last 50 years, some six or seven additional so-called species concepts were proposed (Wheeler and Meier 2000). Are these new species concepts legitimate? To summarize my conclusion, they are not. None of the authors of these new concepts has understood the difference between a species *concept* and a species *taxon*. Instead of new concepts, they have proposed new operational criteria of how to delimit species taxa (see Box 8.2).

A species concept describes the role that the species plays in the living world. Up to now, only two qualifying concepts have been proposed: a species is either a kind, a different thing, and the species definition specifies the criteria according to which species are delimited (typological concept), or a species is considered a reproductive community (biological concept). There is some leeway in the choice of the criteria by which species can be delimited under a given species concept. In Willi Hennig's species definition, the biological species concept was adjusted to the needs of cladification to permit the delimiting of appropriate *clades*. The recognition concept of Hugh Paterson is nothing but a different wording of the BSC. G. G. Simpson's so-called evolutionary species concept contains undefinable criteria and is useless in praxis. And the various so-called phylogenetic species concepts are simply typological prescriptions of how to delimit

Box 8.2   The Three Meanings of Species

The word "species," unfortunately, has different meaning for different people. It leads to great confusion when these differences are not clearly recognized. Most importantly, one must distinguish three different uses of the word species (Bock 1995).

The species concept. I have described how the typological species concept, the concept prevalent among all classic taxonomists, was supplemented (and largely replaced) toward the end of the nineteenth and beginning of the twentieth century by the biological species concept (BSC). Philosophers have referred to typological species as natural kinds. This typological concept is in conflict with the populational nature of species and with their evolutionary potential. Whenever one is in doubt whether to recognize a particular population as a species or not, one can apply the yardstick of the biological species concept—reproductive compatibility. When one is dealing with sympatric populations, the decision is usually clear-cut. However, when allopatric populations are involved, it must be inferred whether they do or do not have the degree of incompatibility one would find in sympatric species. Inevitably such an inference will be somewhat arbitrary. Only two species concepts are or have been in general use, the typological and the biological.

The species taxon. When species are studied over geographical space, it is found that most of them consist of numerous local populations that differ either slightly or more drastically from each other. Such an assemblage of populations distributed in geographic space is a species taxon, as defined by the biological species concept. A species taxon is always multidimensional, whereas the species concept is based on the nondimensional situation. Species taxa that have well-defined subdivisions (subspecies) are called polytypic species.

The species category. This is the rank in the Linnaean hierarchy given to a taxon considered to be a species. The agamospecies recognized by the students of asexual organisms are also ranked as species in the Linnaean hierarchy, even though they do not form populations in the sense of the populations of biological species.

species taxa. None of the putative new species concepts is actually a new concept. They are either rewordings of the two standard concepts or instructions on how to delimit species taxa.

The biological species concept is applicable only to sexually reproducing organisms. Asexual organisms are assigned to agamospecies (see below). In recent years, various other species concepts were proposed, but none of them has been able to take the place of the biological species concept.

The paleontologist G. G. Simpson thought that in paleontology one needed a separate species concept and proposed the evolutionary species concept. However, his definition contains several criteria that cannot be defined. Furthermore, his species definition does not help in the delimitation of species in a phyletic lineage. The phylogenetic species concept is not a concept at all but simply a typological instruction on how to delimit species taxa in a phylogenetic tree. Likewise, the recognition species concept is simply a different formulation of the biological species concept.

THE MEANING OF SPECIES
....

A Darwinian always wants to know why each property of a living organism has evolved. So, he or she asks, "Why are there species? Why are living individuals of sexually reproducing organisms combined into species? Why does the living world not simply consist of independent individuals, each reproducing with some other, somewhat similar individual that he or she encounters?" The reason is obvious, and the study of hybrids between species gives the answer to these questions. Hybrids (particularly in genetic backcrosses) are almost invariably inferior, and often inviable or more or less sterile. This is particularly true for animal hybrids. This demonstrates that genotypes, being well-balanced and harmonious systems, have to be very similar for successful interbreeding. If they are not, as is usually true for the product of species crosses, the hybrid zygotes are apt to be an unbalanced, disharmonious combination of parental genes, resulting in more or less inviable or sterile individuals.

The meaning of species is now quite obvious. The isolating mechanisms of species are devices to protect the integrity of well-balanced, harmonious genotypes. The organization of individuals and populations into species prevents the breakup of well-balanced, successful genotypes as would occur if they crossed with alien, incompatible genotypes, and so it prevents the production of inferior or sterile hybrids. Therefore the integrity of species is maintained by natural selection.

Isolating Mechanisms

But what are these isolating mechanisms? Their definition is: *Isolating mechanisms are biological properties of individual organisms that prevent the interbreeding of populations of different species where they are sympatric.*

This definition makes it quite clear that geographic barriers or any other kinds of purely extrinsic isolation are not isolating mechanisms. For instance, a mountain range that separates two populations that would be able to interbreed if sympatric is not an isolating mechanism. Also, isolating mechanisms, particularly in plants, are often "leaky," that is, they do not prevent the occasional "mistake" that results in the production of a hybrid. However, such occasional hybridism is not sufficiently successful to lead to a general interbreeding and fusion of the two species populations.

Various ways to classify isolating mechanisms have been suggested. The one I have adopted arranges them in the sequence in which these barriers have to be overcome in potential mates (Table 8.3).

Different groups of organisms may have different isolating mechanisms. Mammal and bird species, for instance, are usually kept apart primarily by behavioral incompatibilities. Such species may be fully fertile, as are many species of ducks, yet fail to mate. It is not correct to assume that sterility is the prevailing isolating mechanism. Sterility is apparently more important in plants than in animals, because fertilization in plants is "passive," that is, it is effected by wind, insects, birds, or other extrinsic agents. For this reason, hybrids usually occur more frequently in plants than in higher animals. Yet the production of occasional hybrids leads only rarely to a complete fusion of the two parental species. In plants, however, hybridization may lead through

TABLE 8.3   Classification of Isolating Mechanisms

1. Premating or prezygotic mechanisms: Mechanisms that prevent interspecific matings.
    (a) Potential mates are prevented from meeting (seasonal and habitat isolation)
    (b) Behavioral incompatibilities prevent mating (ethological isolation)
    (c) Copulation attempted but no transfer of sperm takes place (mechanical isolation)

2. Postmating or postzygotic mechanisms: Mechanisms that reduce full success of interspecific crosses.
    (a) Sperm transfer takes place but egg is not fertilized (gametic incompatibility)
    (b) Egg is fertilized but zygote dies (zygotic mortality)
    (c) Zygote develops into an F1 hybrid of reduced viability (hybrid inviability)
    (d) F1 hybrid is fully viable but partially or completely sterile, or produces deficient F2 (hybrid sterility)

allopolyploidy to the production of new species (see Chapter 9). The study of the genetic basis of the various isolating mechanisms is still in its infancy. The number of genes involved to establish reproductive isolation ranges from one, as in the ratio of pheromones in two butterfly species, to the 14 or more that account for the sterility of hybrid males between two closely related species of *Drosophila*.

Hybridization

Hybridization is traditionally defined as the intercrossing of established species. A hybrid is the product of such a cross. Gene exchange among different populations of the same species is frequent (referred to as gene flow), but should not be called hybridization. Rather, hybridization occurs whenever the isolating mechanisms are inefficient ("leaky"). Successful hybridization leads to the transfer ("introgression") of genes of one species into the genome of another species. In some populations, particularly highly inbred ones, this may lead to an enhancement of fitness.

The frequency of hybridization is highly variable. It is rare in most higher animals but frequent in an occasional genus. For instance, there is extensive hybridization among the six species of ground finches (*Geospiza*) on the Galapagos Islands, without apparent loss of fitness. It is also frequent in some families of plants. In spite of the frequency of introgression in such families, hybridization apparently only rarely leads to a fusion of two species and even more rarely to the production of a new species. In plants the doubling of the chromosome number of a sterile species hybrid may lead to the production of a near-fertile allotetraploid species (see Fig. 5.2). In certain groups of vertebrates (reptiles, amphibians, and fishes), species hybrids may shift to parthenogenesis and function as separate species. The F1 hybrid generation may show increased viability ("hybrid vigor") in some species crosses, but this is reversed in the F2 and later generations and in backcrosses. In general, hybrid zones occur when two populations ("species") that have not yet acquired fully effective isolating mechanisms come into secondary contact.

Species Specificity

Even though every individual in a population is uniquely distinct, and every local population is genetically somewhat different from all others, this variability within a species does not mean that members of a species do not share "species-specific" characters. However, these characters are not constant, like an essence, but are always somewhat variable and, more importantly, they have the capacity to evolve in subsequent generations. By far the most important species-specific characters are the isolating mechanisms; others may be ecological properties, such as niche preference.

In spite of numerous diversifying local factors, the continued maintenance of every species is assured by a number of integrating processes. Most important among these is gene flow (see Chapter 5). Equally important is the basically conservative nature of the genotype. The average genotype of a local population is the result of hundreds or thousands of preceding generations of natural selection. Any deviation from this optimum is apt to be selected against by normalizing selection.

However, the selection factors are not the same everywhere in the range of a given species. There is, for instance, the latitudinal change of temperature, and local populations of many species are selected to be best adapted for the local temperature. This results in gradients of characteristics in such a species that parallel the climatic gradients. Such a character gradient is called a *cline*. A cline always refers to a particular character. The geographic variation of a species may involve as many clines as it has geographically variable characters.

Species in Asexual Organisms (Agamospecies)

The equivalent of biological species of sexually reproducing organisms does not exist in asexual organisms. Reproductive communities, such as biopopulations, do not exist in the prokaryotes. Hence, there is considerable uncertainty of how many "species" of bacteria to recognize. Furthermore, bacteria as different as eubacteria and archaebacteria, sometimes classified in two different kingdoms, are known to exchange genes quite frequently by lateral transfer. In such cases, one is forced to fall back on the typological species definition and recognize these species, so-called agamospecies, by the degree of difference.

However, asexual reproduction is also widely found in the eukaryotes. Each asexually reproducing individual belongs to a clone of genetically identical individuals. Whenever a new mutation occurs, it signifies the origin of a new clone. Each clone is a target of selection. Owing to natural selection, many clones are eliminated, producing gaps between bundles of successful clones. If these bundles are separated from each other by sufficiently large gaps, they are considered different species. Speciation in prokaryotes, induced by mutation and the extinction of intermediate clones, is something entirely different from speciation among biological species. *Agamospecies* (asexual lineages), considered to be as different from other groups of such lineages as are biological species taxa, are ranked in the Linnaean hierarchy as species.

In the next chapter I will show how new species can be produced in spite of the various isolating mechanisms to preserve the cohesion of the existing species.

••

CHAPTER 9

In Chapters 5 to 7 I discussed the evolutionary processes that take place in a given population. If these were the only evolutionary processes, the total number of species in the world would always remain the same, even though each species might evolve. And if there was extinction, this would require an answer to the question, Where do the replacement species come from? Lamarck appreciated this problem and solved it by postulating a continuous origin of new species by spontaneous generation. They would be the simplest organisms known to him, but would gradually evolve into higher plants and animals. We now realize that, owing to the current composition of the Earth's atmosphere, such spontaneous generation of new life, having been possible 3.8 billion years ago, can no longer take place. We must look for a different answer.

SPECIATION
••••••••••••••••••

We know that the origin of new species does take place continuously and we must therefore search for the mechanism that produces such a multiplication of species. We want to find out how the millions of existing species originated. This process of the multiplication of species is something entirely different from the phyletic evolution of species in a fossil lineage. But more than that, we also want to know how and why such very different types evolved as bacteria, fungi, giant sequoias, hummingbirds, whales, and anthropoid apes. Indeed, we want to know everything about the evolution of the Earth's amazing organic diversity.

Answers to these questions emerged very slowly. Darwin himself failed to solve the problem of speciation. Even the rediscovery of Mendel's work in 1900 was at first a setback to research in biodiversity, because genetics looked for the answer at the level of the gene. As a result, the leading geneticists, such as T. H. Morgan, H. J. Muller, R. A. Fisher, J. B. S. Haldane, and Sewall Wright, were not able to contribute anything of significance to our understanding of speciation. Their methodology, concentrating on processes taking place in a single gene pool, did not permit them to deal with the issue of biodiversity.

To make progress on speciation it was necessary to adopt an entirely different methodology—the comparison of different populations of a species, that is, the study of geographic variation. And this course was indeed adopted by evolutionary taxonomists, particularly in England, Germany, and Russia. It took more than 60 years after 1859 until the leading specialists of birds, mammals, butterflies, and a few other groups of animals reached an agreement that this geographical approach was the way to solve the problem of speciation. They adopted the theory of *geographical* or *allopatric speciation*, according to which a new species may evolve when a population acquires isolating mechanisms while isolated from its parent population. Unfortunately, the work of these pioneers remained virtually unknown to the mathematical population geneticists. It was not until the 1940s, during the so-called evolutionary synthesis, that the geneticists and the naturalists-taxonomists became acquainted with each other's research and produced a synthesis of their findings (Mayr and Provine 1980).

It was then realized that to understand the origin of biodiversity it was not sufficient to study a single population at different times, so to speak "vertically"; rather, one must compare different contemporary populations of a species with each other. One begins with a comparison of local populations (demes), each consisting of the potentially interbreeding individuals at a given locality. One then studies distinguishable geographical races of a species. These either gradually intergrade with other geographical races of the same species or, when separated by a geographical barrier, they may differ by a clear-cut taxonomic character difference. Indeed some geographical isolates may be so different that it is virtually arbitrary to decide whether to rank them still as geographical subspecies or already as new species. And finally one studies the differences among those species, particularly sympatric ones, that one considers most closely related. By placing these different kinds of populations in a proper sequence, one can reconstruct the pathway of speciation.

Geographical speciation, which seems to be the exclusive mode of speciation among birds and mammals, is the mode that has been most thoroughly investigated (Mayr 1963; Mayr and Diamond 2001). But to more fully consider speciation, one must first review the problem historically.

To understand how one species can give rise to several descendant species, it is necessary to understand what a species is. As shown in Chapter 8, a species taxon is a group of "interbreeding populations reproductively isolated from other such groups." Such a reproductive community of populations is at the same time also different from its ancestors and its descendants and this property led to confusion. Paleontologists, when comparing different temporal populations in a phyletic lineage, often called them different species because they found that they differed from each other, and they used the word speciation for this change. However, such a change in the time dimension does not lead to any increase in the number of species and is best referred to as *phyletic evolution* (Fig. 9.1). When the modern evolutionist speaks of speciation, he or she means the *multiplication* of species, that is, the production of several new species by a single parental species. This is what Darwin had observed on the voyage of the *Beagle* when he concluded that one colonizing South American mockingbird species had produced three different new species of mockingbirds on different islands in the Galapagos. This process is what we now call geographical or allopatric speciation.

THE PROCESS OF ALLOPATRIC SPECIATION

The fundamental question posed by the process of allopatric speciation is: How does the reproductive isolation originate? The answer is found not by looking at the species as a single population, but by expanding our view of the species to a multidimensional species taxon.

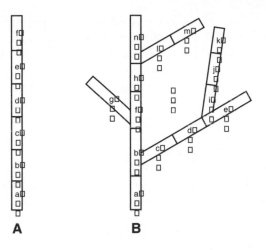

A **B**

FIGURE 9.1

Phyletic evolution vs. speciation. In A (phyletic evolution), after many thousands or millions of years species *a* has evolved into species *f*, but it is still only a single species. In B (speciation), species *a* has given rise to five descendant species (*g, m, n, k, e*) by a process of multiplication of species.

Not all populations of a species taxon are in contiguous contact with each other, actively exchanging genes. Some populations actually are geographically isolated from each other through barriers formed by water, mountains, deserts, or any other kind of terrain unsuitable for this species. These barriers reduce or prevent gene flow in sexually reproducing species and permit each isolated population to evolve independently of the other populations of the parental species. Such a population evolving in isolation is called an incipient species.

What happens in the isolated population? In such a population, numerous genetic processes take place that may differ from similar processes in the parental species. There may be new mutations, certain genes may be lost owing to accidents of sampling, recombination results in the production of a diversity of new phenotypes that are different from those of the parent species, and there may be the occasional immigration of different genes from other populations. More importantly, the isolated population lives in a somewhat different biotic and physical environment from that of the parental species and is therefore exposed to somewhat different selection pressures. In spite of the continuing activity of normalizing selection, the isolated population will gradually be restructured genetically and diverge increasingly from the parental species. If this process continues long enough, the isolated population may eventually become sufficiently different genetically to qualify as a different species. During this process it may acquire new isolating mechanisms that will prevent its interbreeding with the parental population when a change in the nature of the barriers permits the newly evolved species to invade the range of the parental one. When this happens, the incipient species is recognized as a neospecies. The process described here represents geographical or allopatric speciation. What is the fate of the large number of incipient species that are formed all the time? Most of them reunite again with the parental species before having reached the species level or else they become extinct. Only a small fraction of such isolated incipient species completes the speciation process. Actually there are two forms of allopatric speciation.

Dichopatric Speciation

In dichopatric speciation the isolation is caused by the rise of a geographical barrier between two previously contiguous portions of a species (Fig. 9.2). For instance, the flooding of the Bering Strait at the end of the Pleistocene produced a marine barrier between Siberia and Alaska and initiated a divergence in the populations of previously continuous holarctic species, now separated into two portions. Such dichopatric speciation by secondary isolation is most frequent in continental areas. The advance of glaciers at the beginning of each glaciation forced the populations of the retreating species into numerous isolated glacial refuges, where they diverged from each other to a lesser or greater extent. A similar phenomenon seems to have occurred in the tropics, when during arid periods of the Pleistocene the tropical rain forest was reduced to a number of rain forest refugia. Many of the populations in these refugia became new species.

A. Dichopatric (secondary) speciation

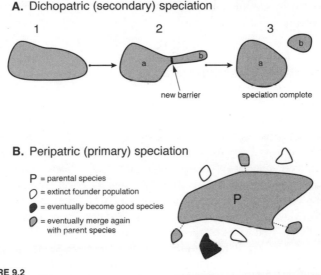

B. Peripatric (primary) speciation

FIGURE 9.2
Two forms of allopatric speciation.

Peripatric Speciation

The isolation in peripatric speciation is caused by the establishment of a founder population beyond the periphery of the present range of a species (Fig. 9.2). This founder population is isolated from the main body of the species by unsuitable terrain and can evolve independently. The importance of peripatric speciation lies in the fact that the founder population is small and genetically impoverished when founded by a single fertilized female or just a few individuals. The gene pool of the new population will be statistically different from the parental gene pool and may facilitate a restructuring of the genotype, particularly the establishment of new epistatic, or intergene, interactions. The founder population is also exposed to the increased selection pressure of an entirely new biotic and abiotic environment. Thus founder populations are potentially in an ideal situation to undertake evolutionary departures into new niches and adaptive zones (Mayr 1954). At the same time, they are exceptionally vulnerable to extinction and to the conservative factor of gene flow. The isolation has to be essentially complete to permit the development of a new species (see Chapter 10).

OTHER KINDS OF SPECIATION

In the 1850s Darwin developed a scheme of speciation based on ecological divergence. He postulated that if different individuals in a population would acquire different niche preferences, they would become different species after many generations. Such speciation would occur without geographic isolation; it would be sympatric speciation. For the next 80 years this was the most widely accepted theory of speciation (Mayr 1992). But this theory was not confirmed in any of the carefully studied cases of speciation in mammals, birds, butterflies, and beetles. In my 1942 *Systematics and the Origin of Species*, I showed that in these groups geographic isolation had been the exclusive mechanism of speciation and not a single case of sympatric speciation had been demonstrated.

Sympatric Speciation

The exclusive occurrence of allopatric speciation in mammals and birds, however, does not refute the possibility of sympatric speciation in other groups of organisms. This has been consistently asserted by entomologists working on insects that specialize on specific host plants (Bush 1994), who have presented evidence for the following scenario. Some individuals of an insect species that is specialized to live on plant species A may colonize plant species B. If the mating of the colonists is restricted to the plant species on which an individual lives, the colonists on B may mate only with other inhabitants of B and may gradually acquire appropriate isolating mechanisms. Such speciation is ordinarily prevented by the continuing colonization of plant B by insects from plant A and by the reverse colonization of plant A by plant B insects. However, there is evidence that in some cases the colonists on B may acquire a preference to mate only with other individuals living on B. This mate preference, then, would be like a barrier between the parental population on A and the colonists on B. In due time this would lead to the sympatric speciation of the colonists on plant B.

Furthermore, there are many cases in freshwater fishes where the occurrence of two or more very closely related species in a rather isolated body of water is best explained by sympatric speciation. For instance, in some small crater lakes in Cameroon two or more very similar species of cichlid fishes coexist that are much more similar to each other than to the ancestral parental cichlid species in the outflow river from the lake. The mechanism by which sympatric speciation took place in this and other similar cases in fishes is a simultaneous preference by females for a certain habitat and for the characteristics of males with the same habitat preference. Such a simultaneous preference was not found in an American cichlid species. Sympatric speciation through the simultaneous acquisition of mate preference (sexual selection) and niche preference has now been demonstrated for several families of freshwater fishes. Hybrids between the two incipient species may be less fit than the parent species. Such cases support the Wallace–Dobzhansky theory of speciation by hybridization. This evidence makes it highly probable that sympatric speciation also occurs in host-specific plant-feeding species of insects, again by the simultaneous preference for niche and mates. However, this does not exclude the possibility that evolution of new host-specific species may also take place by allopatric speciation in founder populations.

Instantaneous Speciation

Through various chromosomal processes an individual may be produced that is instantaneously reproductively isolated from individuals of the parent species. For instance, it occurs quite frequently in plants that a sterile species hybrid AB (with one chromosome set of species A and the other of species B) experiences a doubling of its chromosomes, restoring meiosis and gamete production (AABB). The new polyploid is now a viable species (see Fig. 5.2). By further hybridization and chromosome doubling entire series of polyploids can be produced. What occurs instead in some animals (but has not yet been found in either mammals or birds) is that a sterile species hybrid shifts to parthenogenesis and asexual reproduction. Such cases are known from fishes, amphibians, and reptiles. Again, as in the case of polyploidy, it seems as if such cases of nongeographical speciation are rather rare and likely to be evolutionary dead-ends. Too little is known about reproduction and speciation in the lower animals to say how widespread nongeographical speciation is in those groups.

Parapatric Speciation

According to some evolutionists, a continuous array of populations can break into two separate species along an ecological escarpment. This theory, rejected by most evolutionists, is based on the observation of so-called hybrid belts. These are areas where two rather distinct populations ("species") meet and hybridize. The more widely accepted interpretation of such hybrid belts is that they are areas where two previously isolated incipient species had met in the past but in spite of many differences acquired during the previous isolation had not yet acquired fully effective isolating mechanisms.

Such situations were already known to Darwin. He and Alfred Russel Wallace had an unresolved argument whether or not natural selection could convert a hybrid belt into two full species. Wallace said yes, and was followed in this by Dobzhansky and other modern evolutionists, whereas Darwin said no and was followed in this by H. J. Muller and the present author. A few cases are now known that seem to support the Wallace theory. Usually the hybrid belt is a sink in which the inferior and partially sterile hybrids are steadily eliminated and replaced by immigrants from the adjacent populations of the two parental species. This immigration prevents the selection of balanced intermediates of the two species or individuals with improved isolating mechanisms.

Speciation by Hybridization

Very rarely a hybrid between two species of plants may give rise to a nonpolyploid new species. The rarity of such an occurrence is indicated by the very small number (eight) of cases that have so far been rigorously documented (Rieseberg 1997). They mostly originate in small or peripheral populations. No equivalent cases have so far been found in animals, but some gene exchange (introgressive hybridization) between sympatric species occurs not infrequently in certain groups, for instance, fishes and amphibians, particularly where the habitat was drastically modified by human activities. Fossil plants show that introgressive hybridization may take place between two species for millions of years without affecting the distinctness of the involved species.

Speciation by Distance (Circular Overlaps)

Quite a few cases are known in which a long chain of populations curves around, resulting in an overlap of the ends of this chain. Not surprisingly, the ends of the chain had become so different genetically that they do not interbreed, in other words, they behave toward each other like two different species. Such situations are not in conflict with any principle of Darwinism. Evidently they create, however, a problem in taxonomy. Should such a chain be considered a single species in spite of the sympatry of the ends, or should it be broken into two (or more) species? Much new information favors the second choice. This information comes from a fine-grained analysis of the entire chain. Invariably it shows that the chain only appeared to be continuous, but actually had a number of breaks or remnants of former isolation. When these are recognized as species borders, the "ring" consists of several species and there is no longer any sympatry of two populations of the same species. Two well-analyzed cases are those of the gull *Larus argentatus* (Mayr 1963) and the salamander *Ensatina* (Wake 1997) (Fig. 9.3).

HOW IS THE GENETIC ISOLATION BETWEEN TWO INCIPIENT SPECIES ACQUIRED?

It is rather obvious that the isolating mechanisms must be quite efficient before two incipient species can meet and coexist side by side

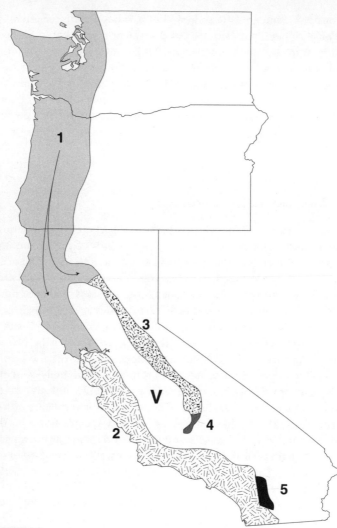

FIGURE 9.3
The "ring species" (circular overlap) of the salamander *Ensatina eschscholtzii*. The species spread from the north (1) around the central valley (V) in two southward streams of populations. One followed the Sierra as subspecies 3, 4, and 5; the other as subspecies 1 and 2 moved along the costal hills. The two portions of the species met in southern California in area 5 and now coexist there without interbreeding.

with only minimal interbreeding. But how can natural selection select for such mechanisms while these populations are geographically isolated from each other? Three possible pathways are usually mentioned, and no complete consensus has yet been reached in this field. Possibly in different cases different pathways were used.

1. The isolating mechanisms evolved in the isolated population as incidental by-products of other, particularly ecological, differences.
2. The differences originated randomly in the isolated populations, a phenomenon well documented by chromosomal differences among isolated populations. In host-specific plant-feeding insects and in parasites, a new host may have been acquired by chance and so provided an isolating mechanism for the new species.
3. By the change of function of characters (see Chapter 10) acquired by sexual selection. It seems that certain color characteristics acquired by males in certain genera of fishes in connection with sexual selection may become behavioral isolating mechanisms when two different populations come into secondary contact.

At one time, particularly when it was believed that mutations would produce new species, there was much discussion of the genetics of speciation and a search for speciation genes. It is now evident that this is not the best way to look at speciation. The definition of the biological species makes it clear that "speciation" means the acquisition of effective isolating mechanisms. This means in turn that the genetics of speciation is the genetics of isolating mechanisms, and also is extremely diversified, because the genetic basis of the various isolating mechanisms is extremely diversified. I do not know of a detailed analysis of the genes involved in any particular instance of speciation, but there are indications that behavioral isolation, as in certain species of cichlid fishes, may be controlled by merely a few genes. By contrast, when whole chromosomes control the reproductive isolation, a large number of genes might be involved. Also, because so many different kinds of isolating mechanisms exist, many different kinds of genes and chromosomes must be involved in speciation. It is not known to what extent, if any, regulatory genes are involved in speciation.

WHAT DETERMINES THE RATE OF SPECIATION?

It was long believed that the rate of speciation was controlled by "mutation pressure." However, there is little evidence to support such a claim. Rather, the rate of speciation is apparently primarily determined by ecological factors. When the range of a species is dissected by geographical and ecological barriers and there is very restricted gene flow in this species, speciation will be rapid and frequent. In island regions or continental regions with insular distribution patterns, there will be much active speciation. On large uniform continents there will be little speciation. Here is a subject that invites much further study. We have good analyses of speciation in certain groups of birds and mammals, but in large groups of animals and plants there is little information on the rate of speciation in various kinds of environments. The most obvious generalization one can make is to say that the less gene flow there is between populations, the more rapidly speciation will occur, all else being equal.

The environment, however, is only one of several factors. There are groups of organisms that speciate rarely or very slowly, an observation for which no ecological explanation has so far been found. This includes the so-called *living fossils*. There are a number of species of plants in eastern North America (including the skunk cabbage) of which populations are also found in a certain area in eastern Asia. These widely separated populations on two different continents are not only morphologically indistinguishable, but apparently also fully fertile with each other, even though they must have been isolated from each other for 6–8 million years. The American botanist Asa Gray called this fact to Darwin's attention (Gray 1963 [1876]). The opposite extreme is represented by the cichlid fishes. For instance, Lake Victoria in East Africa had until recently more than 400 endemic species of cichlids, even though the lake basin was bone dry as recently as 12,000 years ago. Since all species of cichlids in this lake are more closely related to each other than they are to the species in the river flowing out of Lake Victoria, they must have originated within the last 12,000 years. Alas, much of this extraordinary cichlid fauna was recently exterminated by the introduction of a large predatory species, the Nile Bass.

The calculation of average rates of speciation on the basis of the fossil record is apt to lead to biased estimates, because widespread populous species are vastly overrepresented in this record, and they usually have a long life span, and hence a very low rate of speciation. There is a far lower chance that rapidly speciating localized species will be encountered in the fossil record. Considering the enormous range of speciation rates, it is rather questionable whether an "average" rate of speciation has any useful meaning.

· ·

CHAPTER 10

MACROEVOLUTION

When we review evolutionary phenomena, we find that they can be assigned rather readily to two classes. One consists of all events and processes that occur at or below the level of the species, such as the variability of populations, adaptive changes in populations, geographic variation, and speciation. At this level one deals almost exclusively with populational phenomena. This class of phenomena can be referred to as *microevolution*. It was analyzed in Chapters 5–9. The other class refers to processes that occur above the species level, particularly the origin of new higher taxa, the invasion of new adaptive zones, and, correlated with it, often the acquisition of evolutionary novelties such as the wings of birds or the terrestrial adaptations of the tetrapods or warm-bloodedness in birds and mammals. This second class of evolutionary phenomena is referred to as *macroevolution*.

Macroevolution is an autonomous field of evolutionary study. The earlier advances in our understanding of this field were made by paleontologists and systematists. But in recent years molecular biology has made the most important contributions to the understanding of macroevolutionary change, and it continues to make astonishing advances.

From Darwin's day to the present, there has been a heated controversy over whether macroevolution is nothing but an unbroken continuation of microevolution, as Darwin and his followers had claimed, or rather is disconnected from microevolution, as asserted by his opponents, and that it must be explained by a different set of theories.

According to this view, there is a definite discontinuity between the species level and that of the higher taxa.

The reason why this controversy has not been fully settled is because there seems to be an astonishing conflict between theory and observation. According to Darwinian theory, evolution is a populational phenomenon and should therefore be gradual and continuous. This should be true not only for microevolution but also for macroevolution and for the transition between the two. Alas, this seems to be in conflict with observation. Wherever we look at the living biota, whether at the level of the higher taxa or even at that of the species, discontinuities are overwhelmingly frequent. Among living taxa there is no intermediacy between whales and terrestrial mammals, nor between reptiles and either birds or mammals. All 30 phyla of animals are separated from each other by a gap. There seems to be a large gap between the flowering plants (angiosperms) and their nearest relatives. The discontinuities are even more striking in the fossil record. New species usually appear in the fossil record suddenly, not connected with their ancestors by a series of intermediates. Indeed there are rather few cases of continuous series of gradually evolving species.

How can this seeming contradiction be explained? At first sight, there seems to be no method available to explain macroevolutionary phenomena by microevolutionary theories. But should it nevertheless be possible to expand the microevolutionary processes into macroevolutionary ones? And furthermore, can it be shown that macroevolutionary theories and laws are fully consistent with the microevolutionary findings?

The possibility of such an explanation was shown by a number of authors during the evolutionary synthesis, particularly by Rensch and Simpson. They successfully developed Darwinian generalizations about macroevolution without having to analyze any correlated changes in gene frequencies. This approach was consistent with the modern definition of evolution as a change in adaptedness and diversity, rather than as a change in gene frequencies, as suggested by the reductionists. To put it in a nutshell, in order to prove that there is an unbroken continuity between macro- and microevolution, the Darwinians have to demonstrate that seemingly very different "types" are nothing but the end points in a continuous series of evolving populations.

THE GRADUALNESS OF EVOLUTION
···

It is important to emphasize that *all macroevolutionary processes take place in populations and in the genotypes of individuals, and are thus simultaneously microevolutionary processes.* Whenever we study evolutionary change in living populations, we observe such gradualness. Let us consider drug resistance of bacteria. When penicillin was first introduced in the 1940s, it was amazingly effective against many types of bacteria. Any infection, let us say by streptococci or spirochetes, was almost immediately cured. However, bacteria are genetically variable and the most susceptible ones succumbed most rapidly. A few that had acquired by mutation genes that had made them more resistant survived longer and a few still had survived when the treatment stopped. In this manner, the frequency of somewhat resistant strains gradually increased in human populations. At the same time, new mutations and gene transfers occurred that provided even greater resistance. This process of inadvertent selection for greater resistance continued, even though ever stronger dosages of penicillin were applied and the period of treatment was prolonged. Finally, some totally resistant strains evolved. Thus by gradual evolution an almost completely susceptible species of bacteria had evolved into a totally resistant one. Literally hundreds of similar cases have been reported in the medical and agricultural (for pesticide resistance) literature.

Such gradual evolution can be observed wherever one looks. The history of our domestic animals and cultivated plants is a story of gradual evolution even though, in this case, it was effected by artificial selection. Furthermore, fossil-rich geological exposures have recently been found where one can follow a gradual, unbroken series of fossils that demonstrate a gradual change over time.

Even more convincing is the study of geographical speciation (see Chapter 9), in which we can follow how very distinct species by a populational process had gradually diverged from each other. Abundant evidence shows the gradual evolution even of genera. All of this is fully in agreement with Darwinian theory. But this inevitably poses the question, Why is this gradualness not fully reflected in the fossil record?

Darwin already had an answer and, as it turns out, it was indeed

correct. He said that the seeming gaps in the fossil record are an artifact of the haphazard history of the preservation and recovery of fossils. He postulated that the available fossil record was an incredibly
incomplete sampling of the actual formerly existing biota, and that it
was this incompleteness that was responsible for the seeming gaps in
an actually continuous development. All recent research has confirmed Darwin's conclusions. Furthermore, two silent assumptions,
both of them incorrect, have aggravated the difficulties.

SPLITTING VS. BUDDING
••••••••••••••••••••••••••••••

The first assumption was that evolution consists of a splitting of lineages, both of which subsequently diverge from each other at similar
rates. Observation, as well as the theory of speciational evolution (see
below), has shown that this assumption is not necessarily correct. Admittedly, such a splitting of lineages by dichopatric speciation does indeed occur. However, what is apparently far more frequent is that a
new lineage buds off from the parental one by peripatric speciation
and enters a new adaptive zone in which it evolves rapidly, while the
parental lineage remains in its old environment and continues at the
previous slow rate of change.

Let us assume, for instance, that the line leading to birds budded off
one of the various lineages of archosaurs. This new avian lineage, exposed to the powerful selection pressures of the aerial way of life,
changed very rapidly while the parental archosaurian lineage presumably hardly changed at all. That this is a common pattern of evolution
is shown by the fossil record of almost any major taxon, but it is often
overlooked in discussions of theory. The rapid change of the derived
lineage as compared to the slowness of the parental one will undoubtedly be reflected by a gap in the fossil record represented by the period of the rapid changeover from the ancestral condition to the requirements of the new adaptive zone. Remarkably few paleontologists
have given sufficient consideration to the fact that most new evolutionary lineages arise by budding rather than by splitting. And budding is usually achieved very simply by peripatric speciation. Sympatric speciation, likewise, is usually a budding process.

The second misconception held by most students of macroevolution was to think of evolution exclusively as a linear process in the time dimension. When they found a seeming gap in a linear fossil sequence, they assumed either the occurrence of a saltation or an incredible acceleration of evolutionary rate for a short period. Neither assumption fitted the theory of the evolutionary synthesis nor was it supported by credible evidence. Then how can these various discrepancies be explained? What is the explanation of such discontinuities?

Discontinuity

A clearer understanding of evolution was long delayed by a confusion of two meanings of the term discontinuity. One must distinguish between *phenetic discontinuity* and *taxic discontinuity*. A discrete difference among members of the same deme is a phenetic discontinuity. If different members of a mammalian deme have either two or three molars or members of an avian deme have either 12 or 14 tail feathers, it is a phenetic discontinuity. However, if the same difference distinguishes two species taxa from each other, it is a taxic discontinuity. Any discrete difference between two taxa, regardless of taxonomic level, is a taxic discontinuity.

Unfortunately, some typologically thinking evolutionists came to the erroneous conclusion that a phenetic discontinuity would in a single step lead to a taxic discontinuity. In reality a new phenetic discontinuity simply enriches the variation of a deme, producing polymorphism, and it requires a long process of selection to convert a phenetic discontinuity into a discontinuity between two taxa. But when and where is such an individual variation of a deme or group of demes converted into a taxic difference?

SPECIATIONAL EVOLUTION
....

This problem was solved by the students of speciation in living organisms. They showed that species taxa at a given time level not only have the linear dimension of time but also the geographical dimensions of longitude and latitude. Thus they are severely limited both in time and in space. Every species is, so to speak, on all sides surrounded by a gap. Yet it has complete continuity with the parental species from which it descended and the daughter species to which it is giving rise. Furthermore, most species of animals do not consist merely of a single more or less widespread contiguous population but rather are polytypic species, consisting of numerous local populations, many of which, particularly along the periphery of the species range, are more or less isolated from each other. This led to the theory of *speciational evolution* (Mayr 1954), according to which isolated founder populations, established beyond the contiguous species range, may undergo a more or less profound genetic restructuring. This and the subsequent inbreeding of the new population may lead to the production of some unusual new genotypes and of new epistatic balances. Large populations are apparently more inert, less able to break the effects of multiple epistatic interactions than small, genetically impoverished populations. Such small populations are less constrained and able to make greater departures from the ancestral norm. This has been experimentally demonstrated by large and small *Drosophila* populations (see Fig. 6.4). At the same time, the founder population is exposed to new and increased selection pressure owing to the novelty of its new environment. As a result, such a population may rapidly become a different species (see Chapter 9). This theory was also independently arrived at by several botanists (Grant 1963). The chance that such a localized, isolated population, and the new species produced by such peripatric speciation, will be found in the fossil record is, of course, exceedingly small. Even though the continuity of populations during this process of speciational evolution is complete, it will appear in the scanty fossil record as a saltation and has been described as such. This is clearly a misinterpretation, since speciational evolution is at every step a gradual populational process.

Eldredge and Gould (1972) have called this process "evolution by *punctuated equilibria*." They pointed out that if such a new species is successful and becomes effectively adapted to a new niche or adaptive zone, it may subsequently remain unchanged for many hundreds of thousands if not millions of years. Such a *stasis* of a widespread populous species is widely observed in the fossil record.

HOW IMPORTANT IS SPECIATIONAL EVOLUTION?

The theory of speciational evolution was developed not as the result of theoretical considerations but strictly on the basis of actual observations. When studying a series of peripherally isolated populations of a species of birds, the present author noticed that the population that was most peripheral, and that was the product of a sequence of consecutive colonizations, was usually the most different. This observation was fully confirmed and strengthened by the studies of H. L. Carson, K. V. Kaneshiro, and A. R. Templeton on Hawaiian species of *Drosophila*. They showed that colonization of a different island or a different mountain range on the same island might result in a morphologically quite distinct new species, even in a genus with such a stable morphotype as *Drosophila*.

The majority of peripheral isolates, however, differ hardly, if at all, from the parental population. They will have only a rather limited life span and sooner or later become either extinct or merge again with the parental species. However, if we find a somewhat aberrant population in a species, it is almost invariably a far-distant peripheral isolate. This process of speciational evolution has also been referred to as "bottleneck evolution." It may also occur in temporarily highly isolated and in relict populations.

For the new species to be truly successful it must be able to compete with larger, more diversified species. Distributional studies indicate that highly isolated island species in Malaysia and Polynesia are unable to invade the ranges of more widespread species in the West. To become successful when competing with parental and sister species, such founder populations have to increase in size and become more diversified. Such a development is possible for relicts in Pleistocene refuges that, after a change of conditions, can expand their range again and become widespread.

RATES OF EVOLUTIONARY CHANGE

Rates of physical processes, such as chemical reactions or radioactive decay, tend to be constant. This is not at all what we find when we study rates of change in evolution. The evolutionists G. G. Simpson and B. Rensch have been particularly emphatic in calling attention to the great variation in rates of evolution.

Chapter 9 described the high variability of rates of speciation. Equally variable is the rate of simple evolutionary change in phyletic lineages. At one extreme we find the so-called living fossils—certain species of animals and plants that have not visibly changed in more than 100 million years. This includes the horseshoe crab *(Limulus;* Triassic)*, the fairy shrimp *(Triops)*, and the lampshell *(Lingula;* Silurian)*. Equally long lived genera have been found among plants: *Gingko* (dating to the Jurassic), *Araucaria* (probably Triassic), *Equisetum* (mid-Permian), and *Cycas* (*Primo-Cycas*; late Permian).

The complete standstill or stasis of an evolutionary lineage for scores, if not hundreds, of millions of years is very puzzling. How can it be explained? In the case of a living fossil, all the species with which it had been associated 100 or 200 million years ago had either changed drastically since that time or had become extinct. Why did this one species continue to prosper without any changes in its phenotype? Some geneticists thought they had the answer by ascribing it to normalizing selection, which culls all deviations from the optimal genotype. However, normalizing selection is equally active in rapidly evolving lineages. To explain why the underlying basic genotype was so successful in living fossils and other slowly evolving lineages requires a better understanding of development than is so far available.

Not only do species and genera differ from each other in their rate of evolutionary change, but so do entire higher taxa. Paleontologists have shown, for example, that mammals change over time far more rapidly than do bivalve mollusks. In part this difference may be an artifact of the taxonomic method. A bivalve shell has far fewer taxonomic characters than a mammalian skeleton and this discourages a more fine-grained subdivision of bivalve taxa. Yet, even in the most rapidly evolving lineages of animals, the evolutionary change per million years is usually astonishingly low.

We are, of course, fully familiar with the opposite, cases of extraordinarily rapid evolutionary change. This includes the acquisition of immunity to antibiotics in human pathogens and to pesticides among agricultural pests. It is very probable that human populations, living in areas endemic for *Plasmodium falciparum* malaria, have accumulated

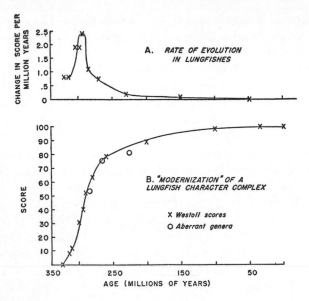

FIGURE 10.1

Rate of acquisition of lungfish characters after the origin of lungfishes. (A) Acquisition of new characters per million years. (B) Rate of approach to the final lungfish body plan per million years. Most of the reconstruction of the body plan of the new taxon takes place in the first 20 percent of its life. *Source*: Simpson, George G. (1953). *The Major Features of Evolution*, Columbia Biological Series No. 17, Columbia University Press: NY.

the sickle cell gene and other blood genes partially resistant to this *Plasmodium* in probably less than one hundred generations.

A phyletic lineage may experience slow and rapid periods of change. A well-known illustration of this phenomenon is the evolution of the lungfishes (Westoll 1949). The major anatomical reconstruction of this class of fishes took place in about 75 million years, while almost no further changes occurred in the ensuing 250 million years (Fig. 10.1). Such a drastic difference between the rates of evolutionary change in young and mature higher taxa is virtually the rule. Bats originated from an insectivorelike ancestor within a few million years, but have hardly changed in basic body plan in the ensuing 40 million years. The origin of whales happened very rapidly, in terms of geological time, compared to the subsequent essential stasis of the new structural type. In all of these cases the lineage had shifted into a new adaptive zone and was for a while exposed to very strong selection pressure to become optimally adapted to the new environment. As soon as the appropriate level of adaptedness had been acquired, the rate of change was reduced drastically. The extreme variability of rates of evolution has been neglected by certain authors and this has led them to errors of interpretation.

How Does One Measure Rates of Evolution?

How long life has existed on Earth was long a complete mystery, and so was the date when the eukaryotes, vertebrates, or insects had originated. But now numerous concrete dates have been established. The oldest fossils (bacteria) are ca. 3,500 million years old, the Cambrian period began 544 million years ago, and the oldest australopithecine fossils are 4.4 million years old. How are these figures obtained?

Geology is the basic source. Many geological strata, particularly volcanic ashbeds or lava flows, contain radioactive minerals, the age of which can be determined by the measurement of their radioactive decay (see Box 2.1). There are now several methods for doing this and the accuracy of the most modern methods is very high.

An entirely different method is available to determine when the common ancestor of two living species lived: the so-called molecular *coalescence method* (see Box 10.1). It is based on the observation that all genes (molecules) change over time at rather uniform rates, and the two lineages derived from a common ancestor become over time more and more different from each other. If the common ancestor is represented by a fossil whose age was determined by geological methods, the average rate of molecular change can be determined accurately (using the molecular clock method). The reliability of this method depends on the constancy of the molecular change. Alas, there are all sorts of irregularities in molecular clock rates and to get reasonably reliable results one must test different materials. Noncoding genes are usually preferable to genes that are subject to changes due to selection. These difficulties are well illustrated by the inferred age of origin of the higher taxa (families and orders) of mammals and birds. The oldest fossils generally fall in the time range of 50–70 million years ago, with no earlier finds, even though there are excellent

Box 10.1   The Coalescence Method of Age Determination

The molecular clock hypothesis states that for all evolutionary lineages there is a relatively constant rate of evolutionary change over time. More specifically, rather than there being a "global" universal rate for all molecules and evolutionary lineages, each molecule, DNA or protein, has a specific rate of evolution. If most mutations are neutral or almost neutral in their selective effects, and if this rate of mutation has not changed over time, then the rate of evolution of a particular molecule should be nearly constant over time permitting us to estimate the age of evolutionary lineages. However, some lineages, have been documented to have, for various reasons, faster rates of evolution than other lineages (e.g. rodents vs. primates). However, leaving this and other caveats aside, if molecules evolve at a constant rate they can be used as "time keepers" to calculate "lineage-specific" divergence times and to estimate the age of the nearest common ancestor of two species.

To use the molecular clock in such a way requires the calibration of its "ticking rate." This can be done through several means such as the fossil record (keeping in mind that the first occurrence of a fossil is always a minimum estimate for the age of this lineage) or through major vicariance events such as plate tectonics. Once the homologous gene A has been sequenced in, e.g., two species and the rate of evolution in this gene is known through prior calibration (let's say 2% per million years) then knowing the percent difference in the DNA sequence of gene A between these two species permits the calculation of the age of their last common ancestor. In this example, if species 1 and 2 differed by 10% in their DNA sequence of gene A, then the common ancestor of these two species would be expected to have lived around 2.5 mya. It would have taken these two lineages this long to both diverge at a rate of 2% per million years to accumulate 10% difference in gene A.

fossil deposits in the crucial period. According to the molecular evidence, these taxa must have originated already in the early Cretaceous, more than 100 million years ago. The cause of this discrepancy is still controversial. Did the molecular clock change its rate?

Neutral Evolution

Molecular genetics has found that mutations frequently occur in which the new allele produces no change in the fitness of the phenotype. Kimura (1983) has called the occurrence of such mutations neutral evolution, and other authors have referred to it as non-Darwinian evolution. Both terms are misleading. Evolution involves the fitness of individuals and populations, not of genes. When a genotype, favored by selection, carries along as hitchhikers a few newly arisen and strictly neutral alleles, it has no influence on evolution. This may be called evolutionary "noise," but it is not evolution. However, Kimura is correct in pointing out that much of the molecular variation of the genotype is due to neutral mutations. Having no effect on the phenotype, they are immune to selection.

SPECIES TURNOVER AND EXTINCTION
....

A striking observation made by paleontologists has been the steady change of biota from one geological period to the next. New species are added to the biota, while old ones disappear because they become extinct. Such extinction does not proceed at the same rate at all times, although a relatively low number of species usually go extinct in any given time span. This *background extinction* has been going on since the beginning of life (Nitecki 1984). The reason for it is that every genotype seems to have limits to its capacity for change and this constraint might prove fatal under certain environmental changes, particularly sudden ones. For example, the needed mutations may have failed to appear when there was either a change in climate or the sudden arrival of a new competitor, predator, or pathogen. Whenever a population is no longer able to reproduce enough offspring to replace losses from natural causes, it will become extinct. No organism is perfect; indeed, as Darwin already emphasized, an organism only has to be good enough to compete successfully with its current competitors. When an emergency arises, there may not be time enough to perfect an adequate genetic restructuring and extinction is the consequence.

This steady extinction of individual species is due to biological causes in almost every case. Furthermore, it is observed in general that the smaller the population size of a species is, the more vulnerable it will be to extinction. However, occasionally a small population seems to be remarkably resistant to extinction.

Actual extinction should not be confused with pseudoextinction. This term, sometimes used by paleontologists, refers to the process by which a species may evolve into a different species and then be given a new name by paleontologists. The ancestral name thus disappears from faunal lists. However, the biological entity involved in this change of names has not become extinct and its seeming disappearance is simply due to a name change.

There are some cases when there was no obvious change in the Earth's environment and yet a major group declined and became extinct. This was perhaps the case with the extinction of the trilobites. Not being able to come up with a better answer, paleontologists have suggested that they succumbed to the competition with the "physiologically superior" newly evolved bivalves. As plausible as this theory appears to be, the evidence for it up to now seems to be rather insufficient. Indeed, some paleontologists now attribute the extinction of the trilobites also to a climatic event.

COMPETITION

The supply of one or several of the resources needed by the population of a species may be limited. In such a case the individuals of this population may be competing with each other (intraspecific competition). Such competition is part of the struggle of existence. It may simply consist of a removal of the limited resources or consist of an actual interference of the competitors with each other. Furthermore, the ecological literature describes numerous examples of competition between individuals of different species. This involves not only similar species, but also competition for seeds between ants and small rodents in the deserts of the southwestern United States. If two species compete too seriously with each other, one of them will be eliminated. Such an occurrence illustrates the competitive exclusion principle, which states that two or more competing species cannot coexist indefinitely when they use exactly the same resources. Such differences may be rather subtle, because cases have been reported in the literature where it has not been possible to find any differences in resource utilization between two coexisting competing species. But such cases are rather rare. Normally competition is a major component of the selection pressure to which the individuals of a population are exposed. And competition between two species for a limited resource often seems to be the reason why one of the two became extinct.

MASS EXTINCTIONS
....

Quite different from the steady extinction of individual species are the so-called *mass extinctions* (Nitecki 1984), during which a large proportion of the biota is exterminated in a very short time on a geological timescale. Mass extinctions are due to physical causes. Most famous among them is the one at the end of the Cretaceous, which involved the extermination of the dinosaurs and of many other marine and terrestrial organisms. For a long time it was a puzzle as to what might have caused this catastrophic extinction, but, as suggested by Walter Alvarez, it is best explained as due to the impact of an asteroid on Earth 65 million years ago. The impact crater of this asteroid has now been discovered at the tip of the Yucatan Peninsula in Central America. The tremendous dust cloud produced by this impact resulted in a drastic drop in terrestrial temperature and in other adverse conditions, producing the extinction of a great proportion of the then existing biota. Although the dinosaurs among the Reptilia became extinct, other reptiles, such as turtles, crocodilians, lizards, and snakes, survived. Some insignificant and probably nocturnal mammals also survived and experienced in the Paleocene and Eocene a spectacular radiation, producing all the orders and many of the families of the now living mammals. The few survivors among the Cretaceous birds seem to have experienced a similarly explosive radiation during the first 20 million years of the Tertiary.

There have been several other mass extinctions since the origin of life on Earth, but those that happened since the origin of the animals

TABLE 10.1   Mass Extinctions

Extinction Event	Age (x 10⁶ years)	Families (%)	Genera (%)	Species (%)
Late Eocene	35.4	—	15	35 +/− 8
End-Cretaceous	65.0	16	47	76 +/− 5
Early–Late Cretaceous (Cenomanian)	90.4	—	26	53 +/− 7
End-Jurassic	145.6	—	21	45 +/− 7.5
Early Jurassic (Pliensbachian)	187.0	—	26	53 +/− 7
End-Triassic	208.0	22	53	80 +/− 4
End-Permian	245.0	51	82	95 +/− 2
Late Devonian	367.0	22	57	83 +/− 4
End-Ordovician	439.0	26	60	85 +/− 3

(metazoans) are best documented (Table 10.1). The most drastic of these other extinctions, apparently even more catastrophic than the Alvarez event, occurred at the end of the Permian and resulted in the estimated extermination of 95 percent of the then existing species. It was apparently not caused by an asteroid impact but by a change of climate or of the chemical composition of the terrestrial atmosphere. There have been three other major mass extinctions (in the Triassic, Devonian, and Ordovician periods), in which 76–85 percent of the then living species became extinct. We are now living in another era of mass extinction caused by humans through the destruction of habitats and the pollution of the environment.

Smaller mass extinctions have happened to specific groups of organisms. During a drought period in the Pliocene (ca. 6 million years ago), the softer C3 grasses in North America were largely replaced by harsh C4 grasses, which have three times as much silica content. Among the browsing horses, all species became extinct except those with the longest teeth.

The Pleistocene extinction of much of the mammalian megafauna of the large continents (including Australia) about 10,000 years ago seems to coincide with a climatic stress period, but also with the appearance of the first efficient human hunters. Presumably both factors contributed to the extinction. That humans were the cause of the extinction of many island faunas (Hawaii, New Zealand, Madagascar, and others) is well documented.

Natural selection, of course, is no protection against mass extinction. Indeed, there is a considerable probability that the successful survival through such an extinction event includes a considerable chance factor. Who, for instance, would have predicted at the beginning of the Cretaceous that the dinosaurs, at that time the most successful group of vertebrates, who occupied such a variety of ecological niches, would be completely exterminated 60 million years later by the Alvarez event? Other previously dominant groups of organisms that also became extinct at the end of the Cretaceous are many marine taxa, such as most nautiloids and the ammonites, both of whom had been previously highly successful organisms. No amount of natural selection succeeded in producing genotypes enabling them to survive.

Background extinction and mass extinction are drastically different in most aspects. Biological causes and natural selection are dominant in background extinction, whereas physical factors and chance are dominant in mass extinction. Species are involved in background extinction, and entire higher taxa in mass extinction. However, certain higher taxa are more susceptible to mass extinction than others. The two kinds of extinction should never be lumped in any statistical analysis of extinction.

MAJOR TRANSITIONS

In spite of its gradualness, macroevolution is characterized by numerous major inventions, which many authors consider to represent decisive steps in the advance of the living world. It begins with the inferred transitions involved in the origin of life and the development of the Prokaryotes. The evolution of life from the Prokaryotes to the most divergent animals and plants is the story of numerous such transitions, such as the rise of the Eukaryotes (with membrane-bonded nucleus, chromosomes, mitosis, meiosis, sex), symbiosis of cellular organelles, multicellularity, gastrulation, segmentation, specialized organs, improved sense organs, elaboration of a central nervous system, parental care, and cultural groups. Almost all of these steps seem to have contributed to the adaptedness of the phyletic lineages in which they occurred (Maynard Smith and Szathmary 1995).

The Origin of Evolutionary Novelties

Some of Darwin's critics readily admitted that an existing structure could be improved by use and disuse or by natural selection, but how could such processes produce an entirely new structure? They would ask, for instance: "How can the origin of wings in birds be explained by natural selection?" Having a small wing, they said, would be of no selective advantage, being useless for flight. Natural selection cannot operate until an already functioning structure is present. Actually, this claim is only a half-truth, because an already existing structure can, by a behavioral shift, assume an additional function that can eventually modify the original structure into an evolutionary novelty. There are two different pathways by which an evolutionary novelty can be acquired: by an intensification of function or by the adoption of an entirely new function (Mayr 1960).

Intensification of Function. In ordinary gradual evolution, most descendant taxa differ from their ancestors only quantitatively. They may be larger, of faster locomotion, more cryptically colored, or differing by some other incremental difference. Nevertheless, the end stages of gradual evolutionary change are often so different from their earliest ancestors that they seem to represent a major saltation. Let us consider the anterior extremities of mammals as an example. Normally, they are adapted for walking, but in moles and other subterranean mammals they are adapted for shoveling earth; in some arboreal mammals, such as monkeys and apes, they are adapted for grasping; in aquatic mammals they become swimming paddles or flukes; and finally in bats they are converted into wings. In all of these cases, except the last one, only a magnification of an existent potentiality is involved. This is what evolutionists refer to as an intensification of function.

Perhaps the most spectacular instance of an intensification of function is presented by the eye. Darwin was puzzled by how such a perfect organ could have evolved gradually. The study of the comparative morphology of organisms has revealed the answer. The simplest, the most primitive stage of the series leading to an eye is a light-sensitive spot on the epidermis. Such a spot is of selective advantage from the very beginning, and any additional modification of the phenotype that enhances the functioning of this light-sensitive spot will be favored by selection. This would include the deposition of pigment around the light-sensitive spot, also any thickening of the epidermis leading up to the development of a lens, of muscles to move the eye, and other accessory structures, but most importantly, of course, the development of a retinalike photosensitive neural tissue.

Photosensitive, eyelike organs have developed in the animal series independently at least 40 times, and all the steps from a light-sensitive spot to the elaborate eyes of vertebrates, cephalopods, and insects are still found in living species of various taxa (Fig. 10.2). They include intermediate stages and refute the claim that the gradual evolution of a complex eye is unthinkable (Salvini-Plawen and Mayr 1977). Most photosensitive organs of the invertebrates lack the perfection of the eyes of vertebrates, cephalopods, and insects, but their origin and subsequent evolution were nevertheless helped by natural selection. As long as a variant was superior, it was favored, with multiple slight advantages reinforcing each other.

Every individual possesses scores, perhaps even hundreds of very slight differences from other members of his or her population. Some observers have felt that these differences would be too slight to be favored by natural selection. This view ignores that many slight advantages can compound and have the effect of one large advantage. Such slight advantages accumulate in the course of generations and thus play an increasing role in evolution. A slight accumulation of pigment and a light-sensitive spot, for example, might not be a special target of selection, but might be favored by survival together with several other equally slight advantages in a phenotype.

The origin of eyes in 40 branches of the evolutionary tree was always considered to be an independent convergent development. Molecular biology has now shown that this is not entirely correct. A regulatory master gene (called *Pax 6*) has recently been discovered that seems to control the development of eyes in the most diverse branches of the tree (see Chapter 5). However, this gene occurs also in taxa

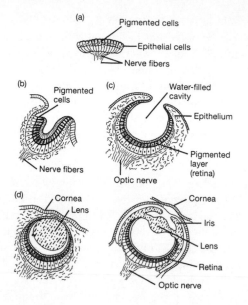

FIGURE 10.2

Stages in the evolution of eyes among molluscs. (a) A pigment spot; (b) a simple pigment cup; (c) the simple optic cup found in abalone; (d) the complex lensed eye of the marine snail and of the octopus. *Source*: *Evolutionary Analysis* 2nd ed. by Freeman/Herron, copyright © 1997. Reprinted by permission of Pearson Education, Inc. Upper Saddle River, NJ.

whose species have no eyes. *Pax 6* is apparently a basic regulatory gene, presumably involving some other functions in the nervous system. Molecular biology has discovered a number of other such basic regulatory genes whose existence in some cases goes back to a time before the major animal phyla had branched. When survival is favored by the acquisition of a new structure or other attribute, selection makes use of all available molecules already present in the genotype.

That a structure like the eye could originate numerous times independently in very different kinds of organisms is not unique in the living world. After photoreceptors had evolved among animals, bioluminescence originated at least 30 times independently among various kinds of organisms. In most cases, essentially similar biochemical mechanisms were used. Virtually scores of similar cases have been discovered in recent years, and they often make use of hidden potentials of the genotype inherited from early ancestors.

Change of Function. Is intensification of function the only way in which complex new organs are acquired? The answer to this question is "No!" There is indeed a second process for such an acquisition, particularly stressed by Darwin, Anton Dohrn, and A. N. Sewertzoff: the acquisition of new organs by the change of function of an existing structure. Such a change requires that this structure is able to perform both the old and the new function simultaneously. For instance, the gliding wing of primitive birds was eventually used also for flapping flight. There are numerous cases of evolutionary novelties that can be explained in this manner. The swimming paddles of *Daphnia* were originally antennae (sense organs) and still function as such, but they are now also used as locomotory structures. Lungs in fishes have been converted into swim bladders and extremities in arthropods have acquired a whole series of new functions. In many cases, what happens is better described as a new ecological role rather than a new function. A structure that is able to adopt a new function is said to be *preadapted* for such a shift. Preadaptation is a purely descriptive term and does not imply any teleological forces.

All the more spectacular origins of new structures or habits in the history of organisms were due to a change of ecological role. Such shifts splendidly illustrate the opportunism of evolution. As stated in Jacob's (1977) *principle of tinkering*, any existing structure may be used for a new purpose.

A change of function may also play a role in some cases of speciation. It is possible, particularly in the case of sympatric speciation, that a factor favored by sexual selection takes on the new role of a behavioral isolating mechanism.

Any change of function event simulates a saltation, yet it is actually a gradual populational change. It affects at first only one individual within a population and becomes evolutionarily significant only if it is favored by natural selection and spreads gradually to the other individuals of the population and then to the other populations of the species. Hence, even evolution by change of function is a gradual process.

ADAPTIVE RADIATION
••••••••••••••••••••••••••

Whenever a species acquires a new capacity, it acquires, so to speak, the key to a different niche or adaptive zone in nature. The branch of reptiles that invented the feather and subsequently the capacity to fly conquered an enormous adaptive zone. As a result, birds now have about 9,800 species as compared to only 4,800 species for all the mammals and 7,150 species of living reptiles. The structural type that we call "insect" is particularly successful, having given rise to several million species. However, all attempts by birds to conquer water have been only mildly successful. There are about 150 species of ducklike birds, and a few grebes (20), auks (21), and loons (4), while the penguins, the most water adapted of all aquatic birds, have only 15 species—thus only two percent of all species of birds are aquatic. A considerable number of species of mammals have succeeded in becoming leaf-eaters, but only a few birds, most successfully the hoatzin, succeeded in conquering that niche. No amphibian succeeded in adapting to salt water.

The History of Life: A Story of Adaptive Radiation

The success of a phyletic lineage to establish itself in numerous different niches and adaptive zones is called *adaptive radiation*. It is conspicuous in most of the higher taxa of organisms. Reptiles, without abandoning their basic structure, evolved into crocodilians, turtles, lizards, snakes, ichthyosaurs, and pterosaurs; mammals produced mice, monkeys, bats, and whales; and birds evolved into the niches of hawks, storks, songbirds, ostriches, hummingbirds, and penguins. Each of these groups has carved out its own suite of niches in nature, without any major change in the ancestral structural type.

Actually, the entire ascent of life can be presented as an adaptive radiation in the time dimension. From the beginning of replicating molecules to the formation of membrane-bounded cells, the formation of chromosomes, the origin of nucleated eukaryotes, the formation of multicellular organisms, the rise of endothermy, and the evolution of a large and highly complex central nervous system, each of these steps permitted the utilization of a different set of environmental resources, that is, the occupation of a different adaptive zone.

Disparity

The diversity of the living world takes many forms. It may express itself purely quantitatively as in the large colonies of ants and termites, or in the number of species in a family, like the weevils among the beetles (and the order of beetles as a whole), and of course in the enormous biomass of prokaryotes. But diversity may also express itself in the degrees of difference, the number of strikingly different types of organisms. And here evolution has produced a real surprise. In the rise of the metazoans (animals), one would expect that soon after their appearance in the fossil record they would consist of a series of rather similar orders that would become increasingly more dissimilar to each other in the course of time. Yet the facts are astonishingly different from this assumption! When the metazoans appeared as fossils about 550 million years ago (admittedly they must have already existed for ca. 200 million years), they included four to seven bizarre body plans that soon became extinct. All the other Cambrian phyla survived, and what is quite unexpected, without a major revolution of the basic body plan. If we look at individual phyla, the same situation is encountered. The living classes of arthropods are already found in the Cambrian with the same body plans. But again there are a handful of strange types of arthropods in the Cambrian that do not exist today. I agree with those who conclude from this evidence that the variety of realized body plans was greater in the Cambrian than it is now. Furthermore, no fundamentally new body plan has originated in the 500 million years since the Cambrian.

The solution to this puzzling problem will have to be supplied by developmental biology. Development in the recent phyla is rigidly canalized by *hox* genes and numerous other regulatory genes. There are indications that this regulatory system has considerably tightened since the Cambrian. Hence, at the time of the origin of the metazoans, the constraining power of the regulatory system was apparently still very rudimentary. Seemingly rather minor mutations might have produced totally novel structures. This "freedom of construction" was lost as the regulatory machinery was increasingly perfected and now, hundreds of millions of years later, different feeding types of cichlid fishes can still originate, but all are still cichlid fishes. To say that the body plans of the living fauna display the same disparity as that of the Cambrian is simply not true. And yet the contrast between the innovativeness of the Cambrian fauna and the conservativeness of the body plans of the living fauna is no longer an insolvable puzzle when the recent findings of developmental molecular biology are duly considered.

COEVOLUTION
....

Whenever two kinds of organisms interact with each other, let us say a predator and its prey, or a host and its parasite, or a flowering plant and a pollinator, each will exert a selection pressure on the other. The result is that they will coevolve. For instance, the prey may develop better escape mechanisms that force the predator to improve its attack capacity. Much of the process of evolution occurs through such *coevolution*.

The pollinators of the flowers of plants, whether they are butterflies, other insects, birds, or bats, are adapted to the flowers of their host plants and these flowers, in turn, evolve in such a way as to make the pollination more successful. Darwin conducted a fascinating study of the adaptations of orchids for pollination. All cases of symbiosis or mutualism found in nature are subject to such coevolution due to natural selection.

Plant species protect themselves against herbivores by the production of all sorts of toxic chemicals, such as alkaloids, which make them unpalatable to potential herbivores. The herbivores then develop detoxifying enzymes to overcome this problem. In response, the plants develop new chemicals for their protection. The herbivores then have to develop again the appropriate detoxifying enzymes to combat these new toxins. Such a series of back and forth interactions has been referred to as an "evolutionary arms race," and there are an almost infinite number of such arms races among organisms. Marine snails, for instance, protect themselves against snail-eating crabs by evolving stronger shells as well as all sorts of structural elaborations of the shell that make it more difficult for the crabs to crush them. The crabs, in turn, develop stronger claws, which induces the snails to grow even tougher shells, and so on.

Obviously it is not the best evolutionary strategy for a pathogen to wipe out its host. Indeed, there should be a premium on the evolution of less virulent strains. It is sometimes possible to observe such an evolution taking place. When, for instance, the myxomatosis virus was introduced into Australia to control the escalating population of rabbits, the most virulent strains of this virus killed their host rabbits so quickly that there was no time for the virus to be transmitted to another rabbit. As a result, most of the highly virulent strains became extinct. Rabbits attacked by less virulent strains survived longer and provided the source for infecting other rabbits. Eventually, much less virulent strains of the virus evolved that killed only a certain percentage of the rabbits while most survived. At the same time, the most susceptible rabbits were killed off and populations of rabbits evolved that were less susceptible to the myxomatosis virus.

Most European infectious diseases currently exist in a similar steady state. Over many millennia, the European populations have become somewhat resistant to these human diseases and mortality is relatively low. This was not the case, however, with foreign populations that first came in contact with the Europeans after 1492. All over the world, but particularly in the Americas, the native populations were ravaged by epidemics caused by European infectious diseases, particularly smallpox. The native population of the Americas, which was estimated to have been 60 million when Columbus first landed in the Bahamas, had crashed to 5 million only 20 years later. These diseases were so deadly because the Native Americans had not coevolved with them. They were left defenseless when the pathogens spread through their populations.

Internal parasites, such as cestodes, trematodes, and nematodes, tend to become gradually host specific after they have colonized a new host, and from that point on they evolve together with their host. Whenever the host splits into two species, the parasite in due time will do the same. As a result, it is sometimes possible to construct a phylogenetic tree of the parasite that parallels that of the host. There are exceptions, because once in a while a parasite may be able to jump to an entirely different lineage of hosts. What is true for internal parasites is equally true for external ones, such as lice, feather lice (Mallophaga), and fleas.

SYMBIOSIS

In the discussion of evolution, not nearly enough attention is paid to the overwhelming role of symbiosis. *Symbiosis* is the collaboration of two different kinds of organisms in producing a system of reciprocal helpfulness. Lichen, a system consisting of a fungus and an alga, is an oft-cited case of symbiosis. It is apparently widespread among bacteria, resulting in the evolution of entire bacterial communities, for instance, among soil bacteria, in which different kinds of bacteria produce different metabolites useful to other species.

All insects that feed on plants and plant juices have intracellular symbionts that produce enzymes needed for the digestion of the plant material. Blood-sucking insects likewise often have intracellular symbionts facilitating the digestion of blood.

The most important event in the history of life on Earth, the production of the first eukaryotes, was apparently initiated by the symbiosis between a eubacterium and an archaebacterium, leading eventually to the formation of a chimaera between these two kinds of bacteria. Additional events led to the incorporation of symbiotic purple bacteria in the new eukaryote to form the mitochondria, and in plants to the symbiotic incorporation of cyanobacteria into the cell to become chloroplasts. Other cellular organelles are also symbionts (Margulis 1981; Margulis and Fester 1991; Sapp 1994).

EVOLUTIONARY PROGRESS

Evolution means directional change. Since the beginning of life on Earth and the rise of the first prokaryotes (bacteria) 3,500 million years ago, organisms have become far more diversified and complex. A whale, a chimpanzee, and a giant sequoia are surely very different from a bacterium. How can this change be characterized?

The answer most frequently given is that current life is simply more complex. On the whole this is indeed true, but it is not universally true. Many phyletic lineages demonstrate simplifying trends, and this is particularly true for various kinds of specialists such as cave animals and parasites. But surely, it will be said, evolution shows progress. Are not vertebrates and angiosperms (flowering plants) more highly evolved, more progressive, than "lower" animals and plants, and bacteria? We have already analyzed this claim and shown how difficult it is to apply the designations "higher" and "lower." In fact, the prokaryotes, as a whole, seem to be as successful as the eukaryotes. Yet, every step in evolution, generation after generation, that eventually led to rodents, whales, grasses, and sequoias took place, so to speak, under the control of natural selection. Does not this lead by necessity to a steady improvement, generation after generation, of every phyletic lineage? The answer is "No," because most evolutionary changes are dictated by the need to cope with current temporary changes of the physical and biotic environment. Hence, considering also the enormous frequency of extinction and the occurrence of regressive evolution, it is inevitable that one must reject the notion of universal progress in evolution. However, a different answer can perhaps be given when one looks at single lineages at particular moments of their evolution. There are a considerable number of phyletic lines that one could well call progressive during the period of their greatest flowering.

DOES SELECTION LEAD TO PROGRESS AND ULTIMATELY TO PERFECTION?

In the eighteenth century it was widely believed that the world was perfectly designed by God, and that even where such perfection had not yet been achieved, he had instituted laws that would ultimately lead to it. This belief reflected not only the thinking of natural theology but also the optimism of the Enlightenment, as well as the teleological thinking (finalism) that was so widespread in that period. Lamarck's theory of evolution, for instance, postulated a steady rise toward perfection. Modern evolutionists reject the idea that evolution is able ultimately to produce perfection. Yet most of them believe that some sort of evolutionary progress has occurred since the beginning of life. The gradual change over time from bacteria to unicellular eukaryotes, and finally to flowering plants and higher animals, has often been referred to as progressive evolution. Such terminology has been used particularly often with reference to man as the end stage of a series leading from reptiles through primitive mammals to placentals and finally to monkeys, apes, and hominids. At one time the idea was almost universally held that man was the culmination of Creation and that anything was progressive that led in the direction of man's perfection.

Doesn't the series from bacterium to man indeed document progress? If so, how can such seemingly progressive change be explained? In recent years a number of books were published debating the existence or validity of evolutionary progress. There is great dissension on this question because the word "progress" has so many different meanings. For instance, those who adopt teleological thinking will argue that progress is due to a built-in drive or striving toward perfection. Darwin rejected such a causation and so do modern Darwinians, and indeed no genetic mechanism was ever found that would control such a drive. However, one can also define progress purely empirically as the achievement of something that is somehow better, more efficient, and more successful than what preceded it. The terms "higher" and "lower" have also been criticized. For the modern Darwinian it is not a value judgment, but "higher" means more recent in geological time or higher on the phylogenetic tree. But is any organism "better" by being higher up on the phylogenetic tree? Progress, it is claimed, is indicated by greater complexity, more advanced division of labor among organs, better utilization of the resources of the environment, and better all-around adaptation. This may be true to some extent, but the skull of a mammal or bird is not nearly as complex as that of their early fish ancestors.

Critics of the concept of progress have pointed out that in some ways bacteria are at least as successful as vertebrates or insects, and therefore why should vertebrates be considered progressive over prokaryotes? The decision as to who is right depends largely on what one considers to be progress.

If one looks at the evolutionary series, one cannot deny that some recently evolved taxa have adaptations that were particularly successful for survival. Warm-bloodedness, for instance, permits an organism to cope more successfully with climate and weather fluctuations than is possible for ectotherms. A large brain and extended parental care permit the development of culture and its transmission from generation to generation (see below). Each of these advances has been the result of natural selection, with the survivor having had an advantage over the nonsurvivors. In this descriptive sense, evolution was clearly progressive in certain phylogenetic lineages. It was as progressive as the development of the modern motor car from such early types as Ford's Model T. Each year the manufacturers of motor cars adopted new innovations and these were then exposed to the selection pressure of the market. Many models with certain innovations were eliminated; the successful ones formed the basis for the next level of innovation. As a result, the cars improved from year to year, becoming safer, faster, more durable, and more economical. Surely the modern car represents progress. If we consider a modern car as representing progress over the Model T Ford, we are equally justified to call the human species progressive compared to lower eukaryotes and prokaryotes. It all depends on how we interpret the word "progressive." However, Darwinian progress is never teleological.

Many definitions of evolutionary progress have been offered. I particularly like one that emphasizes its adaptationist nature: Progress is "a tendency of lineages to improve cumulatively their adaptive fit to their particular way of life, by increasing the number of features which combine together in adaptive complexes" (Richard Dawkins, *Evolution* 51(1997): 1016). For other definitions and descriptions of progress, see Nitecki (1988).

The incorporation of symbiotic prokaryotes evidently was a highly progressive step by the first protists, resulting in the immensely successful empire of the eukaryotes. Other progressive steps have often been cited: multicellularity, the development of highly specialized structures and organs, endothermy, highly developed parental care, and the acquisition of a large, efficient central nervous system. The "inventors" of each new progressive step were also highly successful and this contributed to their ecological dominance. Indeed, the gist of every selection event is to favor individuals that have succeeded in finding a progressive answer to current problems. The summation of all of these steps is evolutionary progress.

To continue my analogy, the development of the motor car by no means displaced all other modes of transportation. Walking, the horse, the bicycle, the railroad, they all still coexist with the motor car, all being used under certain circumstances. Nor did the invention of the airplane make the railroad or the motor car obsolete. It is the same with organic evolution. Rather primitive prokaryotes still survive more than 3 billion years after their first appearance on Earth. Fish still dominate the oceans and, except for humans, rodents are more successful in most environments than primates. Also, as shown by cave inhabitants and by parasites, evolution is often retrogressive. However, it is quite legitimate to refer to the series of steps from the prokaryotes to eukaryotes, vertebrates, mammals, primates, and man as progressive. Each step in this progression was the result of successful natural selection. The survivors of this selection process have been proven to be superior to those that were eliminated. The end product of all successful so-called arms races can be considered to be examples of progress.

BIOSPHERE AND EVOLUTIONARY PROGRESS

Most accounts of the history of life on Earth are written as if the environment had been constant, but actually it was not. In particular, there was a drastic change in the composition of the atmosphere. At the time when life originated (ca. 3.8 billion years ago), the atmosphere was reducing, consisting presumably of some mixture of methane (CH_4), ammonia (NH_3), molecular hydrogen (H_2), and water vapor (H_2O). There was hardly any free oxygen, and whatever was produced by cyanobacteria disappeared quickly in various sinks, among which the oxidation of iron to iron oxide was the most conspicuous. This led to the deposit of the so-called banded iron formation. The supply of oxidizable iron in the world's oceans was exhausted ca. 2 billion years ago. The continuing production of free oxygen by cyanobacteria quickly converted the anoxic atmosphere into an oxygen-rich atmosphere and this contributed to the evolution of a rich fauna of multicellular animals. It is believed that the so-called Cambrian "explosion" of new animal types was assisted by the simultaneous enrichment of the atmosphere by oxygen.

The evolutionary changes of the biota during the last 550 million years have greatly affected the composition of the atmosphere. Most important have been the conquest of land by plants (beginning about 450 million years ago), the development of rich angiosperm forests with their capacity to consume CO_2, and the evolution of detritus-consuming bacteria.

Vernadsky (1926) was the first to point out the ongoing coevolution between oxygen-producing and oxygen-consuming organisms, as well as the changes in the biota in response to gradual as well as cataclysmic changes in the environment, such as mass extinctions. Organisms can respond to changes of the environment only if they can quickly produce the appropriate variants needed by natural selection. If they do not, they become extinct. Oxygen is not the only element in very active interchange with organisms. Others include calcium (chalk, limestone, corals, shells) and carbon (coal, oil). Changes in the world's climate have of course also had great evolutionary effect, particularly glaciations and correlated changes in the course of ocean currents, particularly around Antarctica.

HOW CAN WE EXPLAIN TRENDS IN EVOLUTION?

Often when paleontologists compare related organisms in succeeding strata they discover "trends." For example, the later descendants may be increasingly larger than their ancestors. This trend toward increased size is very widespread among animal lineages and is known as Cope's Law. A trend may be described as a directional change in a feature in a phyletic lineage or in a group of related lineages. For instance, in a study of horse evolution during the Tertiary, it was discovered that there was a tendency for a reduction in the number of toes, so that the modern horse has only a single one of its original five toes. At the same time, in certain lineages of horses there was a tendency in the molar teeth to become higher and to continue growing throughout life. This is referred to as hypsodonty. Trends such as these were discovered in ammonites, trilobites, and virtually all types of invertebrates. An increase in brain size, not only in primates, is a widespread trend in the evolution of Tertiary mammals. A trend in one specially favored character (e.g., hypsodonty in horses) may result in trends in various correlated characters. In other words, a particular trend may be nothing but the by-product of a trend in a different character, such as body size.

Some paleontologists were puzzled by the seeming linearity of some of these trends. Selection, they claimed, is far too haphazard a process to account for such linearity. This argument, however, overlooks that any evolutionary change in a series of organisms is subject to severe constraints, as shown by the constraints on an increase in the size of the teeth of a horse exerted by the size of the body. There is, for example, a severe constraint on body size in flying organisms, which is why the flying taxa of vertebrates (bats, birds, pterosaurs) are only a fraction of the size of their largest terrestrial relatives. Furthermore, almost all trends are not consistently linear, but change their direction sooner or later, sometimes repeatedly, and they may even totally reverse their direction.

In the days when teleological thinking was widespread, trends were interpreted as evidence for intrinsic tendencies or drives. This was used as the major evidence for a rather popular school of evolutionists who believed in teleological orthogenesis (see Chapter 4). The almost lawlike progression in some of these trends was interpreted by this school as being incompatible with Darwin's natural selection. Subsequent research, however, has shown that there is no such conflict. No support for the existence of intrinsic evolutionary trends was ever found and trends can be explained quite confidently by the Darwinian model with due consideration of constraints. It is now quite evident that all observed evolutionary trends can be fully explained as being the result of natural selection.

Correlated Evolution

An organism is a carefully balanced, harmonious system, no part of which can change without having an effect on other parts. Let us consider the increase in the size of teeth in horses. This change requires a larger jaw, and in turn a larger skull. To carry the larger skull, the entire neck has to be reconstructed. The larger new skull has an effect on the rest of the body and in particular on locomotion. This means that in order to acquire larger teeth virtually the whole horse must to some extent be reconstructed. This has been confirmed by a careful study of the anatomy of hypsodont horses. Also, since the whole horse had to be reconstructed, the change could occur only gradually and slowly over many thousands of generations. Many lineages of horses with low molar teeth failed to come up with the required genetic variation for hypsodonty and became extinct.

The shift from the quadrupedal locomotion of a lizardlike reptile to bipedalism and flight in birds initiated a considerable restructuring of the body plan: a compacting of the whole body to have a better center of gravity, the development of a more efficient four-chambered heart, restructuring of the respiratory tract (lungs and air sacs), endothermy, improved vision, and an enlarged central nervous system. The acquisition of all of these adaptations was a matter of necessity. Details, however, are often dictated by constraints and the availability of genetic variation.

Sometimes the development of one aspect of the phenotype may have unexpected consequences for other parts of the body. This is well illustrated by evolution among the reptiles. Two major subdivisions of the Reptilia are recognized: the Synapsida, with one temporal skull opening, and the Diapsida, with two openings. The turtles, without any temporal opening, were believed to be an old group that had originated before the development of any temporal openings. Molecular analysis, however, has shown that the turtles are diapsids, related among living reptiles to the crocodilians. Apparently they lost the skull openings during the acquisition of the carapace as part of a general reduction of all openings to the outside. This, incidentally, also shows how drastically a taxonomic character may change its value during evolution.

Complexity

Many early evolutionists were convinced that evolution advanced steadily toward ever greater complexity. Indeed, the prokaryotes, which represented life on Earth for more than 1 billion years, are far less complex than the eukaryotes, which evolved subsequently. But among the prokaryotes there is no indication of ever increasing complexity in the long period of their existence. Nor does one find any evidence for such a trend among the eukaryotes. To be sure, multicellular organisms are, on the whole, more complex than the protists, but at the same time numerous evolutionary lineages are found among both plants and animals that evolved from complexity to greater simplicity. The skull of a mammal, for instance, is far less complex than that of its placoderm ancestors. Wherever we look, we find simplifying trends as well as trends toward greater complexity. Parasites are, on the whole, notorious for their many physical and physiological simplifications. All theories that postulated the existence in all organisms of an intrinsic trend toward greater complexity have been thoroughly refuted. There is no justification in considering greater complexity to be an indication of evolutionary progress.

MOSAIC EVOLUTION

Organisms never evolve as types; there is always a greater selection pressure on some properties than on others, and these attributes then evolve faster than the others. In the evolution of man, for instance, there are enzymes and other proteins that have not changed in six or more million years, and are therefore still identical with those of chimpanzees or even earlier primate ancestors. Other primate properties of hominids have changed drastically, with the central nervous system changing the most. The Australian *Platypus* has hair and suckles its young with milk and has other characteristics of primitive mammals, but lays eggs, like reptiles, and has some "dead-end" specializations, like a poison spur and a duckbill. This uneven rate of evolution of different properties of an organism is called *mosaic evolution*, and it may create difficulties for classification. The first species of a new branch of a phylogenetic tree will have acquired a single derived key character but may agree in everything else with its sister species. Darwinian taxonomists usually classify such a species with its sister species with which it agrees in most of its characters. A Hennigian cladist, however, may assign it to a new clade.

The fact that the evolution of different components of the phenotype of an organism may to some extent be independent of each other provides great flexibility for evolving organisms. To successfully enter a new adaptive zone, an organism might have to change only a limited component of its phenotype. This is well illustrated by *Archaeopteryx*, which in many respects (e.g., teeth, tail) is still a reptile, even though it has the feathers, wings, eyes, and brain of a bird. Mosaic evolution is even more strikingly demonstrated by the highly different rates of evolution of different proteins and other molecules.

Not knowing how to explain mosaic evolution, geneticists long ignored it. Now a theory of "gene modules" has been proposed, in which the concerted action of certain groups of genes ("modules") has been postulated. Such modules can, to some extent, evolve rather independently of each other.

PLURALISTIC SOLUTIONS
....

Evolution is an opportunistic process. Whenever there is an opportunity to outcompete a competitor or to enter a new niche, selection will make use of any property of the phenotype to succeed in this endeavor. Several different solutions are usually available for any challenge by the environment.

Flying was invented by vertebrates three different times, but the wing of each flying taxon—birds, pterosaurs, and bats—is different. Even more different are the wings of different kinds of insects, for instance, dragonflies, butterflies, and beetles, although all of them seem to be derived from a single ancestral flying type.

Pluralism is characteristic of all aspects of the evolutionary process. Genetic variation is replenished in most eukaryote species by sexual reproduction (recombination), whereas in the prokaryotes it is replenished by unilateral gene transfer. Reproductive isolation is effected in most higher animals by prezygotic isolating mechanisms (e.g., behavior), and in others by chromosomal incompatibilities, sterility, or other postzygotic factors. Speciation usually occurs for geographic reasons in terrestrial vertebrates, but it is sympatric in certain groups of fishes and perhaps in plant-host-specific groups of insects. There is a very reduced amount of gene flow in some species, while others disperse so easily that the entire species is virtually panmictic. Furthermore, some families have many actively speciating genera, while others have only a few old monotypic genera.

In view of this rampant pluralism, at the level of both micro- and macroevolution, it is advisable to exercise great caution when applying the findings for one group of organisms uncritically to others. Findings made in one group of organisms do not necessarily refute different findings made in another group.

CONVERGENT EVOLUTION

Convergent evolution is a phenomenon that convincingly illustrates the power of natural selection. The same ecological niche or adaptive zone is often filled on different continents by exceedingly similar, but entirely unrelated organisms. The opportunity provided by the same adaptive zone results in the evolution of similarly adapted phenotypes. This process is called *convergence*. The most famous case is that of the Australian marsupials. These indigenous mammals have, in the absence of placental mammals, produced types analogous to placental mammals in the northern continents. The northern wolf is matched by the Tasmanian wolf, the placental mole by the marsupial mole, the flying squirrel by the marsupial phalanger, and there are other less close analogs: a mouse, a badger (wombat), an anteater. (Fig. 10.3). Species adapted to subterranean life (and convergently similar) have independently evolved in four different orders of mammals and among the rodents in eight different families (Nevo 1999). Such cases of convergent evolution are not exceptional, but are actually quite widespread. To mention a few others: the American and the African porcupine, the New World vultures (Cathartidae, related to storks) and the Old World vultures (Accipitridae, related to hawks), and the nectar-feeding birds—hummingbirds (Trochilidae) in the Americas, sunbirds (Nectariniidae) in Africa and southern Asia, honeyeaters (Meliphagidae) in Australia, and honeycreepers (Drepanididae) on Hawaii (Fig. 10.4). Any knowledgeable zoologist would be able to list several pages of such cases of convergent evolution.

Convergent evolution of vertebrates in the ocean produced sharks, porpoises (mammals), and the extinct ichthyosaurs (reptiles). Convergent developments have occurred in many animal taxa, but they also occur among plants. The various kinds of cactus in America are paral-

FIGURE 10.3

Convergent evolution of Australian marsupials (right) and placental mammals (left) on other continents. Each pair is similar in form and lifestyle. *Source: A View of Life* by Salvador E. Luria et al. Copyright © 1981 Benjamin Cummings. Reprinted by permission of Pearson Education, Inc.

leled by analogs among the Euphorbiaceae of Africa (Fig. 10.5). Convergence illustrates beautifully how selection is able to make use of the intrinsic variability of organisms to engineer adapted types for almost any kind of environmental niche.

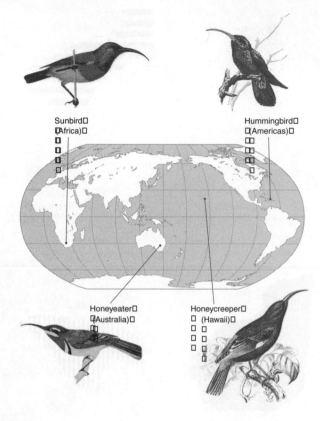

FIGURE 10.4

Independent evolution of nectar-feeding adaptations in four songbird families: sunbird (Nectariniidae), hummingbird (Trochilidae), honeyeater (Meliphagidae), and honeycreeper (Hawaiian finches, Drepanididae). *Sources*: Honeycreeper (Hawaii), Wilson, S.B. and Evans, A.H. (1890–1899). *Aves Hawaiienses: The Birds of the Sandwich Islands*; Honeyeater (Australia), Serventy, D.L. and Whittell, H.M. (1962). *Birds of Western Australia* (3rd ed.) Paterson Brokensha: Perth; Sunbird (Africa), Newman, K. (1996). *Newman's Birds of Southern Africa: The Green Edition*. University Press of Florida: Gainesville, FL. Reprinted by permission of Struik Publishers of Cape Town, South Africa and Kenneth Newman; Hummingbird (Americas), James Bond (1974) *Field Guide to the Birds of the West Indies*. HarperCollins Publishers.

POLYPHYLY AND PARALLELOPHYLY
••

In the pre-Darwinian days of classification, convergent groups were often combined into a single taxon owing to their similarity. Such a

A

B

FIGURE 10.5
Parallel evolution of similar arid country adaptations in (a) American cactuses and (b) African euphorbs. (From Starr et al. 1992.) *Source*: Photographs copyright © 1992, Edward S. Ross. Reprinted by permission.

taxonomic assignment is known as *polyphyly*. Recognition of such a polyphyletic taxon was in conflict with Darwin's demand that every taxon should be monophyletic, that is, should consist exclusively of descendants of the nearest common ancestor. Darwinian taxonomists broke up such polyphyletic taxa and placed the parts with their nearest relatives. The combination of whales and fishes was such a polyphyletic taxon that was later rejected.

Convergence must be carefully distinguished from *parallelophyly*, which designates the independent emergence of the same character in two related lineages descended from the nearest common ancestor (Fig. 10.6). For instance, stalked eyes occur independently and irregu-

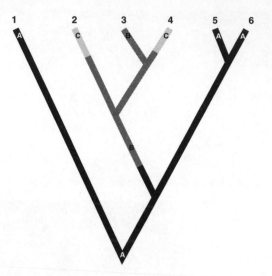

FIGURE 10.6
Parallelophyly. The independent evolution of similar phenotypes (2, 4) owing to the inheritance of the same propensity in the common ancestral genotype (3).

larly in various lineages of acalypteran flies, because all of these lineages have inherited from their common ancestor the genotypic capacity for the production of such eyes. But this propensity has been realized in only some of the lineages. Many if not most cases of *homoplasy* are caused by such parallelophyly. In a reconstruction of phylogeny, not only the phenotype must be considered but also the ancestral genotype and its phenotypic potential.

A Case Study: The Origin of Birds

The greatest current controversy in phylogeny will perhaps be settled by invoking parallelophyly; it concerns the origin of birds. There is no argument over the conclusion that birds derived from the archosaurian lineage of the diapsid reptiles. But when this happened is the argument. As far back as the 1860s, T. H. Huxley called attention to the remarkable similarity of the avian skeleton to that of certain reptiles and concluded that the birds had descended from dinosaurs. Later, other authors postulated a much earlier origin, but recently the dinosaur origin has been proclaimed by the cladists with such vigor that at present it seems to be the most widely accepted explanation of the origin of birds. Indeed, the similarity of the pelvis and legs between birds and certain bipedal dinosaurs is astonishingly close (see Fig. 3.6).

However, the arguments of their opponents are also very persuasive. The fossil chronology seems to be in conflict with the dinosaur theory. The particular bipedal dinosaurs that are most birdlike occurred in the later Cretaceous, some 70–100 million years ago, while *Archaeopteryx*, the oldest known fossil bird, lived 145 million years ago. *Archaeopteryx* has so many advanced avian characters that the origin of birds must be placed considerably earlier than the late Jurassic, perhaps in the Triassic, but no birdlike dinosaurs are known from that period. Furthermore, the digits in the dinosaurian hand are 2, 3, 4 while in the avian hand they are 1, 2, 3. Also, the anterior extremities of the birdlike dinosaurs are very much reduced and in no way preadapted to become wings. It is quite inconceivable how they could have possibly shifted to flight. These are only a few of the numerous facts in conflict with a Cretaceous origin of birds from a dinosaurian ancestry. The argument will probably not be fully settled until more Triassic fossils are found.

ARE THERE LAWS OF EVOLUTION?
....

This is a question that physicists and philosophers like to ask. To answer it, one first needs to decide what one means by the word "law." The kind of laws characteristic of the physical sciences, which can be stated in mathematical terms and have no exceptions, are sometimes also encountered in functional biology. Mathematical generalizations can often be applied to biological phenomena, like the Hardy-Weinberg equilibrium relating to the distribution of alleles in populations. By contrast, all so-called evolutionary laws are contingent generalizations, and thus not equivalent to the laws of physics. Evolutionary "laws," such as Dollo's Law of the irreversibility of evolution or Cope's Law of an evolutionary increase in body size, are empirical generalizations, with numerous exceptions, and are quite fundamentally different from the universal laws of physics. Empirical generalizations are useful for ordering observations and in the search for causal factors. Rensch (1947) made a particularly helpful contribution to this subject in pointing out that evolutionary "laws" are greatly restricted in time and place and therefore do not satisfy the traditional definitions of scientific laws.

CHANCE OR NECESSITY?

For years there has been a rather heated controversy over whether chance (contingency) or *necessity* (adaptation) is the dominant factor in evolution. Enthusiastic Darwinians tended to ascribe every aspect of a living organism to adaptation. They argued that in every generation there is a drastic culling of each population, sparing on the average only two of the hundreds, thousands, or in some cases even millions of offspring of each set of parents. Only the most perfectly adapted individuals, they would claim, could pass through this ruthless process of elimination. Those who uphold adaptation as the dominant force in evolution have indeed a strong argument.

Unfortunately, some of the strict adaptationists forgot that natural selection is a two-step process. To be sure, selection for adaptedness is paramount at the second step, but this is preceded by a first step—the production of the variation that provides the material for the selection process, and here stochastic processes (chance, contingency) are dominant. And it is this randomness of variation that is responsible for the enormous, often quite bizarre diversity of the living world. Let us consider two cases. The first is the enormous diversity of the unicellular eukaryotes ("protists"). Margulis and Schwartz (1998) recognize in this kingdom no fewer than 36 phyla of mostly unicellular eukaryotes, many of them parasitic. These include such utterly diverse organisms as amoebas, radiolarians, foraminifera, sporozoans, *Plasmodium*, zooflagellates, ciliates, green algae, brown algae, dinoflagellates, diatoms, *Euglena*, slime molds, and chytridiomycota, to mention just a few of the better-known ones. Another specialist recognized as many as 80 phyla. Many of them are strikingly different from each other, and for some of them, it is still argued whether they should not rather be classified with fungi, plants, or animals. Does it

really require that many different body plans for unicellular eukaryotes to be well adapted?

The diversity among the multicellular organisms is even more astonishing. Not only do we have multicellular "protists" like the brown
algae, but the differences between and within the three rich multicellular kingdoms, the fungi, plants, and animals, are even more overwhelming. Did they need all of these differences in order to be well
adapted? Let us look at the bizarre types in the Burgess shale fauna.
One cannot escape the suspicion that many of them were due to muta-
tional accidents that were not eliminated by selection. Indeed, I sometimes wonder whether the elimination process is not sometimes a
good deal more permissive than is usually assumed. Furthermore, one
must not forget that chance always plays a considerable role even at
the second step of evolution, that of survival and reproduction. And
not all aspects of adaptedness are tested in every generation.

Or let us look at the 35 or so living phyla of animals. They are the
survivors of the 60 or more body plans that existed in the early Cambrian. When one studies their differences, one does not get the im-
pression that they are necessities. Many or even most of their unique
characteristics may have had their origin in a developmental accident
that was tolerated by selection, while the seeming failure of those that
became extinct may have been the result of a chance event (like the
Alvarez asteroid extinction event). S. J. Gould (1989) made such con-
tingencies a major theme in *Wonderful Life*, and I have come to the
conclusion that here he may be largely right.

One can conclude from these observations that evolution is neither
merely a series of accidents nor a deterministic movement toward
ever more perfect adaptation. To be sure, evolution is in part an adaptive process, because natural selection operates in every generation.
The principle of adaptationism has been adopted so widely by Darwinians because it is such a heuristic methodology. To question what
the adaptive properties might be for every attribute of an organism
leads almost inevitably to a deeper understanding. However, every attribute is ultimately the product of variation, and this variation is
largely a product of chance. Many authors seem to have a problem in
comprehending the virtually simultaneous actions of two seemingly
opposing causations, chance and necessity. But this is precisely the
power of the Darwinian process.

Can we also apply this conclusion to man? Some of the most enthusiastic promoters of the principle of contingency have claimed "Man is nothing but an accident." This conclusion is, of course, in complete conflict with the teachings of most religions, which consider man the pinnacle of Creation or the end point of a long drive toward perfection. The success—at least in terms of population growth and expanding range—of mankind in the last 500 years would seem to demonstrate how well man is adapted. On the other hand, if the making of man had been a deterministic process, why did it take 3,800 million years to produce? The species *Homo sapiens* is only about a quarter million years old, and prior to that time our ancestors were in no way outstanding within the animal kingdom. No one could have predicted that a defenseless, slow-moving biped should become the pinnacle of Creation. But one of the australopithecine populations somehow acquired the brain power to survive by its wits. One can hardly avoid considering this more or less of an accident, but it wasn't a pure accident because every step in the change from an australopithecine to *Homo sapiens* was furthered by natural selection.

IV
HUMAN EVOLUTION

CHAPTER 11

···

HOW DID MANKIND EVOLVE?

Man has always been considered something entirely different from the rest of Creation. This is stated in the Bible, and the philosophers, from Plato to Descartes to Kant, entirely agreed with this conclusion. To be sure, some eighteenth-century philosophers placed man on the *scala naturae*, but this had no influence whatsoever on the views of the average person. For most people, man was the crowning of Creation and differed from all animals in multiple ways, particularly by the possession of a rational soul. Therefore, it came as a terrible shock to the Victorian age when Darwin, following his theory of common descent, incorporated the human species into the animal kingdom as a descendant of primate ancestors. Even though Darwin himself was at first rather cautious in how he expressed himself, some of his followers, such as Huxley (1863) and Haeckel (1866), were quite emphatic in proclaiming apes to be man's ancestors. Darwin himself eventually gave a full account of his views on man's evolution in his *Descent of Man* (1871).

The visible similarity between man and apes, of course, had not escaped the attention of earlier naturalists. Indeed, Linnaeus had included the chimpanzee in the genus *Homo*. Nonetheless, not only theologians and philosophers, but in fact virtually everybody else had simply ignored this obvious similarity. Lamarck's account of the evolution of man was likewise ignored. However, Darwin's new theory of common descent, in which all living organisms are derived from common ancestors, made the recognition of man's primate origin inevitable.

WHAT ARE THE PRIMATES?
······································

The Primates are an order of mammals consisting of the prosimians (lemurs and lorises), tarsiers, New World monkeys, Old World monkeys, and apes (Table 11.1). They are not very closely related to any other mammalian order, their nearest relatives evidently being the flying lemurs *(Galeopithecus)* and tree shrews (Scandentia). The earliest primate fossils are of late Cretaceous age.

The Old World monkeys gave rise to the apes 33–24 million years ago (mya). The fossil monkey *Aegyptopithecus* (late Oligocene) already had some anthropoid (apelike) characteristics. *Proconsul* (23–15 mya) of eastern Africa was clearly an ape, ancestral to man and the African apes, but unfortunately there are no African anthropoid fossils from 6 to 13.5 mya (Fig. 11.1).

The living apes consist of two groups, the African apes (the gorilla, chimpanzees, and man) and the Asian apes (gibbons and the orang). There is a definite gap between these two groups; the branching apparently took place some 12–15 mya.

TABLE 11.1   Classification of Primates

Order Primates
    Suborder Prosimii
        Infraorder Lemuriformes (lemurs)
        Infraorder Lorisiformes (galagos, lorises)
    Suborder Tarsiiformes (tarsiers)
    Suborder Anthropoidea
        Infraorder Platyrrhini (New World monkeys)
        Infraorder Catarrhini (Old World monkeys)
            Superfamily Hominoidea (apes)
                Family Hylobatidae (gibbons)
                Family Hominidae
                    Subfamily Ponginae (Pongo orang)
                    Subfamily Homininae (African apes, humans)

These groups of primates were originally recognized on the basis of morphological differences. The validity of these groups and their relationship to each other have been confirmed in recent years by molecular characteristics.

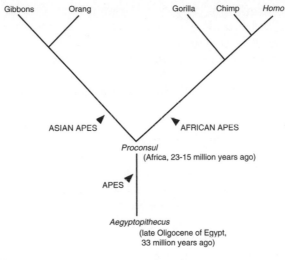

FIGURE 11.1
Phylogeny of the apes.

WHAT EVIDENCE SUPPORTS THE
PRIMATE ORIGIN OF MAN?
••••••••••••••••••••••••••••••••

No well-informed person any longer questions the descent of man from primates and more specifically from apes. The evidence for this conclusion is simply too overwhelming; it consists primarily of three kinds of facts.

Anatomical Evidence.   Right down to minor details, humans agree in all anatomical structures with the African apes, particularly the chimpanzee. R. Owen once thought he had found a real difference in the structure of the brain, but T. H. Huxley refuted this claim; the difference is only quantitative, not qualitative. The same turned out to be true for later similar endeavors. The few strictly human characteristics are differences in the proportion of arms and legs, the mobility of the thumb, body hair, skin pigmentation, and size of the central nervous system, particularly the forebrain.

Fossil Evidence.   In 1859, when Darwin published his daring findings, no fossils were known that would have supported the gradual transition from a chimpanzeelike ancestor to modern man. Although even today no fossils have yet been found from the period between 5 and 8 mya, during which the branching event took place, numerous fossils dating from 5 mya to the present document the nature of the intermediate stages (see below) between chimpanzees and humans.

Molecular Evolution.   One of the great achievements of molecular biology has been to show that macromolecules evolve exactly like visible structural characteristics. Hence a comparison of the human macromolecules with those of apes might shed light on human evolution, and so it does. Indeed, it shows that human molecules are more similar to those of chimpanzees than to any other organism, and furthermore that the African apes are more similar to man than they are to any other kind of primates. The similarity is so great that certain enzymes and other proteins of man and chimpanzee are still virtually identical, for instance, hemoglobin. Others differ slightly, but the difference is less than that between chimpanzees and monkeys.

One can summarize this voluminous anatomical, fossil, and molecular evidence by stating that the very close relationship between man and chimpanzee and other apes has now been convincingly documented. It would be quite irrational to question this overwhelming evidence.

WHEN DID THE HOMINID LINEAGE BRANCH OFF FROM THAT LEADING TO THE CHIMPANZEE?

In other words, how old is the hominid lineage? In the days when man was still considered entirely different from any animal, the branching point was placed way back in time, perhaps at the beginning of the Tertiary, some 50 mya. When more fossils and more and more similarities between man and the African apes were discovered, more recent dates were successively accepted. For quite some time, a date of 16 million years was widely accepted. When a study of the proteins

and DNA differences finally permitted the establishment of a molecular clock, the findings suggested that the branching point was as recent as 5 to 8 mya. Subsequent findings by a number of different
methods support this date. By these methods it was also established
that the branching point between man and the chimpanzee appears to
be more recent than that between chimpanzee and gorilla. That is, the
evidence now suggests that the chimpanzees are our nearest relatives,
and that they are more closely related to man than to gorillas.

WHAT DOES THE FOSSIL RECORD TELL US?

Only a few hominid fossils were discovered prior to 1924 and all represented the most recent stages in hominization, or the rise of the
genus *Homo*. These finds were made in Europe, Java, and China. This
led to the widespread assumption that man had originated somewhere
in Asia, and large expeditions ventured into Central Asia to look for
early fossils. Alas, they did not succeed. Even though some perceptive
authors had already pointed out that an African origin was much
more likely, owing to man's relationship to the chimpanzee and gorilla, it was only in 1924 that the first fossil hominid was discovered in
Africa *(Australopithecus africanus)*. Since then, numerous additional
finds have been made in Africa, indeed it is only in Africa that fossil
hominids older than 2 million years have been found. There is now
no longer any doubt that Africa was the cradle of mankind.

The Ascent of Fossil Man

It has been customary in the anthropological literature to tell the story
of fossil man in the form of a chronology of the discoveries. It usually
began with Neanderthals (1849, 1856), went on to *Homo erectus* (1891
[Java], 1927 [China]), and then to the African discoveries (from 1924
on). For an evolutionist, however, it makes more sense to begin with
the earliest fossils and gradually move on to the discoveries of the geo-
logically more recent ones. This is the approach that I adopt.

The chimpanzee lineage, well after its separation from the hominid line, split into two allopatric species. One is the widespread chimpanzee *(Pan troglodytes)*, ranging across all of Africa from the west to the east, and the other one is the bonobo *(Pan paniscus)*, who is restricted to the forests on the western bank of the Congo River in Central Africa. This river separates the two species. In some of its behavior, the bonobo seems to be more similar to humans than is the chimpanzee, but this does not mean that the bonobo was our ancestor. The branching event between chimpanzee and bonobo took place only a few million years ago, long after the hominid and chimpanzee lines had split.

How to Reconstruct the Path from Ape to Man?

One of the tasks of paleoanthropology is to reconstruct the sequence of the changes from ape to man. The early students of fossil man who attempted such reconstructions had been trained as anatomists and were highly qualified to describe these changes. However, conceptually they were not equally well prepared for this task. They were typologists, thinking in terms of a change from "Ape" to "Man." What they wanted to find were the steps in the gradual change of the type ape into the type man. They also had an almost teleological belief in a linear trend "toward more perfection," a progressive trend culminating in *Homo sapiens*.

Alas, the reconstruction of the steps of hominization proved to be very difficult. First of all, the first fossils that were found were the most recent ones. So the path of reconstruction was not from ape to man but from man back to ape. More disturbingly, it turned out to be quite impossible to establish the hoped for smooth continuity. This, of course, was largely due to the incompleteness of the fossil record, but not entirely so, and this is what was so disturbing. As we shall see (see below for details), some fossil types were relatively common and widespread, such as *Australopithecus africanus*, *A. afarensis* and *Homo erectus*, but they were seemingly separated by discontinuities from their nearest ancestors and descendants. This is particularly true for the break between *Australopithecus* and *Homo*.

WHAT IS THE ACTUAL FOSSIL EVIDENCE?
···

Unfortunately, no hominid fossils—nor such of a fossil chimpanzee—are as yet known for the period between 6 and 13 mya. Thus there is no documentation of the branching event between the hominid and the chimpanzee lineages. To make matters worse, most hominid fossils are extremely incomplete. They may consist of part of a mandible, or the upper part of a skull without face and teeth, or only part of the extremities. Subjectivity is inevitable in the reconstruction of the missing parts. From the beginnings of human paleontology there has been a tendency to compare every fossil with *Homo sapiens*. A fossil (or particular parts of it) was then considered "advanced" or primitive ("apelike"). These comparisons showed that hominid evolution tended to be highly "mosaic." A very *Homo*-like dentition may be associated with rather apelike extremities, and other rather incongruous combinations were also found.

A general text on evolution like this one cannot present the cons and pros of all interpretations of the controversial hominid finds (and virtually all of them are somewhat controversial!). This would be totally bewildering for the nonspecialist reader. What I have done, and will surely be widely criticized for, is to select among the numerous interpretations that one that seemed to me the most likely correct one. The reader must realize that the assignment of each fossil in this treatment is provisional. Any new find may drastically change the situation. Proposals such as the tentative placement of *Homo habilis* with the australopithecines or the immigration of *Homo* into eastern Africa from elsewhere in Africa are particularly vulnerable. It is important in this bewildering situation not to take anything for granted. Tattersall and Schwartz (2000) provide a most helpful account of the variation of hominid fossils. Anthropologists coming into hominid classification from anatomy must remember that taxonomic species names like *afarensis, erectus,* and *habilis* do not designate types but rather variable populations and groups of populations.

Our incomplete knowledge of the fossil hominids is highlighted by the fact that no less than six new species of fossil hominids were described in the seven years since 1994. No one has yet attempted to properly place them in a new hominid phylogenetic tree. What portion of the differences among the various fossils is due to geographic variation cannot be determined on the basis of the few scrappy remains.

STAGES IN HOMINIZATION
••••••••••••••••••••••••••••••

Yet, as far as the general trend in human evolution is concerned, the fossil record is of considerable assistance. By making use of the interpretations of numerous authors, but relying particularly on Stanley (1996) and Wrangham (2001), I am developing a sequence of historical narratives that reconstruct the various steps in the history of the change from ape to man. The resulting picture is entirely based on inferences and any part of it may be refuted at any time. But developing a cohesive story is far more instructive than merely compiling a list of unconnected facts. The most important certainty that has emerged from recent studies is that *Homo sapiens* is the end product of two major ecological shifts (habitat preference) of our hominid ancestors. As a result, one can distinguish three stages of hominization:

The Rain Forest Stage	Chimpanzee
The Tree Savanna Stage	*Australopithecus*
The Bush Savanna Stage	*Homo*

The Chimpanzee Stage.   Rain forest apes move from tree to tree commonly by brachiation. Their main foods are soft fruits and other soft plant material (leaves, stems, etc.). The small brain and great sexual dimorphism are diagnostic for the apes. They spend most of their life in trees and there is no selection pressure for bipedalism.

The Australopithecine Stage.   Around 5–8 million years ago, some species of chimpanzeelike ape succeeded in establishing founder populations in the belt of tree savanna surrounding the rain forest. A huge area of Africa at that time was apparently occupied by the tree savanna and these colonists evolved into the australopithecines. They were apparently immensely successful and presumably occurred wherever there were tree savannas in Africa, even though at present their fossils have been found only in eastern Africa from Ethiopia to Tanzania and in South Africa. There is a single find in Chad (central Africa).

In order to become adapted to this new habitat, these apes had to change remarkably little. At this time the trees were more often some distance from each other and the apes had to adopt bipedal locomotion, but they essentially remained arboricolous and usually slept in tree nests, like other apes. A shift to bipedal locomotion may not be as difficult for a primate as is sometimes believed. I have seen South American spider monkeys move considerable distances bipedally in the Phoenix (Arizona) Zoo. The only other adaptation apes had to acquire was longer and harder teeth, since they had to include tougher plant material in their diet as there was probably a shortage of soft tropical fruit in this more arid habitat. Some anthropologists believe that they discovered the edibility of underground storage organs of plants, such as tubers, rhizomes, and corms, which occur in more arid habitats. Lions, cheetahs, wild dogs, and other carnivores that outrun their prey were rare or absent in the tree savanna, and trees were always available for escape from predators. As a result, the australopithecines had no need to change most of their ancestral chimpanzee characters, such as small size, large sexual dimorphism (males being about 50 percent larger than females), a small brain, long arms, and short legs.

There are two well-documented gracile species of australopithecines: *A. afarensis* in eastern Africa from Ethiopia to Tanzania (3.9–3.0 mya) and *A. africanus* in southern Africa (3.0–2.4 mya) (Fig. 11.2). Both have a small brain of about 430–485 cc. Although they are allospecies, *A. africanus* is younger and more similar to *Homo* except in the proportions of its extremities. Considering that chimpanzees were already quite proficient in tool use, one would expect the same from the australopithecines, but so far no flaked stone tools of theirs have been discovered. Whatever tools they may have made from wood, plant fiber, and animal skins have not survived. There is no reason not to assume that the australopithecines lived in tree savannas throughout Africa.

Australopithecus was largely a vegetarian. Its incisors were larger than those of man, as were its molars, which are considerably smaller in chimps.

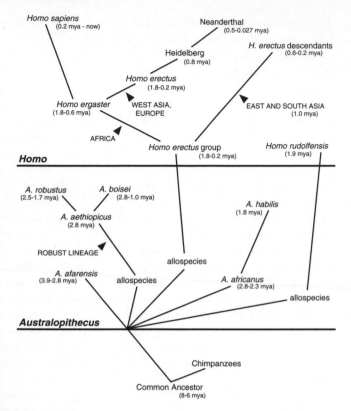

FIGURE 11.2

Very tentative suggestion of the hominid phylogeny. The given dates of their occurrence are particularly prone to revision. Hominids described after 1990 are not included.

Although bipedal, *Australopithecus* apparently still lived mostly arboreally and much of its body structure, like the length of the arms, was quite different from that of modern man. According to Stanley (1996), this means that the females could not carry their infants in their arms (which were needed for climbing), and the young infant had to be able to hang on to its mother, as infant apes do. Likewise, the infant had to be born as advanced as, for instance, a young chimpanzee.

There are rather few genera of primates (e.g., *Cercopithecus*) in either the Old or New World in which two different species coexist in the same area. But this is the case in the australopithecines. In the same region in southern Africa where the gracile species *A. africanus* lived, *A. robustus*, a member of a robust lineage, also lived. And in eastern Africa the robust *A. boisei* is found from about 3.5 to 3.0 mya,together with the gracile *A. afarensis*, and with *Homo* from 2.4 to 1.9 mya. An even older robust species, *A. aethiopicus*, existed around 3.8 mya, but may not be separable from *A. boisei*. Although the robust australopithecines appear to have been very powerful, all indications are that they were peaceful vegetarians. Basically they have the same body structure as the gracile australopithecines, but some authors place the robust australopithecines in the genus *Paranthropus*.

The gracile *Australopithecus* populations lived from 3.8 to 2.4 mya. In their body size and smallness of the brain they were apes. What is most noteworthy, however, is that they did not change very much in this whole 1.5-million-year-long period; it was a period of stasis. To be sure there were differences between the southern African *A. africanus* and the eastern African *A. afarensis*, who lived at somewhat different times, but the differences might also be attributed to geographic variation induced by climatic and other environmental conditions. There was no approach toward the characters of *Homo* over this long period.

WERE THE AUSTRALOPITHECINES APES OR HUMANS?
••

This question was the subject of a heated controversy when *A. africanus* was discovered in 1924. The outcome depended, of course, on the evaluation of the characters by which *Australopithecus* differs from *Pan* and *Homo*. Ever since *Homo* was acknowledged to be an ape, its upright posture and bipedal locomotion were considered characteristic human properties, and since *Australopithecus* shared these properties with *Homo*, the australopithecines were ranked with the humans. In part of the nineteenth century and most of the twentieth, bipedalism was considered a very important character. It was argued that the upright posture freed arms and hands for other roles, in particular for the making and using of tools. This, in turn, required brain activity and was the main reason for the increase of human brain size.

Bipedalism thus was considered the most important stepping stone in hominization.

   This chain of reasoning is no longer convincing. The australopithecines were bipedal for more than 2 million years and yet over this whole period there was no significant change in the size of their brain. Tool use, likewise, has been downgraded in importance owing to the discovery of extensive tool use by chimpanzees and of rudimentary tool use by corvids and other animals. Furthermore, except for bipedalism and some tooth characters, the australopithecines shared almost all their other characters with the chimpanzees. And, what is surely more important, they had none of the most typical *Homo* characters. They lacked a large brain, they did not produce flaked stone tools, they still had the strong sexual dimorphism of apes, they had long arms and short legs, and their body size was small. Also we must distinguish between two forms of bipedalism, that of the arboricolous australopithecines and that of the exclusively terrestrial humans. It is probably correct to claim that in the aggregate of their characteristics the australopithecines were closer to chimpanzees than to *Homo*. Indeed, the step from the *Australopithecus* apelike stage to the *Homo* stage was clearly the most important event in the history of hominization.

THE CONQUEST OF THE BUSH SAVANNA

Human history always seems to have been vitally affected by the environment. Beginning about 2.5 mya, the climate in tropical Africa began to deteriorate, correlated with the arrival of the ice age in the Northern Hemisphere. As it became more arid, the trees in the tree savanna suffered and gradually more and more of them died and the environment slowly shifted to a bush savanna. This deprived the australopithecines of their retreat to safety, for in a treeless savanna they were completely defenseless. They were threatened by lions, leopards, hyenas, and wild dogs, all of whom could run faster than they. They had no weapons such as horns or powerful canines, nor the strength to wrestle with any of their potential enemies successfully. Inevitably, most australopithecines perished in the hundreds of thousands of years of this vegetational turnover. There were two exceptions. Some tree savannas survived in especially favorable places and here some australopithecines also survived for a while, such as *A. habilis* and the two robust species *(Paranthropus)*.

More important for human history, however, is the fact that some australopithecine populations survived by using their wits to invent successful defense mechanisms. What these were can only be speculated about. The survivors could have thrown rocks, or used primitive weapons made from wood and other plant material. They might have used long poles like some chimpanzees from western Africa, swung thorn branches, and perhaps even used noise-making instruments like drums. But surely fire was their best defense and, not being able to sleep in tree nests, they most likely slept at campsites protected by fire. They also were the first humans to make flaked stone tools, and it is possible that they used sharper flakes to construct lances. The fact is that these descendants of the australopithecines, now evolving into *Homo*, survived and eventually prospered. The arboricolous bipedalism of the australopithecines evolved into the terrestrial bipedalism of *Homo*.

This shift was the most fundamental one in all of hominid history. It was a far greater change than the habitat shift from rain forest to tree savanna and resulted in the evolution of a series of important diagnostic characters of the new genus *Homo*. Brain size rose quickly and more than doubled in *H. erectus*. Sexual dimorphism declined from a 50 percent to a 15 percent higher weight of the males. The teeth, particularly the molars, became much smaller. The arms shortened and the legs lengthened. Early *Homo* seems to have relied on fire not only for protection but apparently also for cooking. The reduction in tooth size in *Homo* has traditionally been ascribed to an increased reliance on meat in their diet. But Wrangham et al. (2001) believe that the softening of tough plant material by cooking was a more important cause. Almost everything in this scenario is controversial. The date when fire was tamed is particularly uncertain, and some of the early recorded dates have turned out to be misinterpretations. And if fire was as important for the evolution of *Homo* as it now seems, it must have been relied on already by the earliest *Homo*, but this has not yet been documented.

THE ORIGIN OF *HOMO*
....

The evolution of *Homo* is documented by fossil discoveries, even though rather skimpily. Around 2 million years ago, a very different kind of hominid appeared suddenly in eastern Africa. It was first described as *Homo habilis*, but soon it was realized that the specimens described under this name were too variable to belong to a single species, and the larger-brained specimens were separated and described as *H. rudolfensis*. As more specimens were found, the interpretation changed drastically. The name *habilis* was restricted to the smaller specimens. The brains of the *"Homo" habilis* specimens measured only 450, 500, and 600 cc, thus widely overlapping *Australopithecus*, while the brain of *H. rudolfensis* measured from 700 to 900 cc, being strikingly larger (Table 11.2). *Homo rudolfensis* also differed from *Australopithecus* in other characters; it had shorter arms and longer legs, its cheek teeth were smaller, and its incisors were larger. The stone tools originally ascribed to *habilis* are now credited to *H. rudolfensis*, and *"Homo" habilis* is now considered a late species of *Australopithecus*. The reason why the whole situation is so puzzling is that *H. rudolfensis* does not seem to have descended from any known species of *Australopithecus* in eastern or southern Africa. Rather, it seems to have invaded eastern Africa from somewhere else in Africa. Surely, there must have been australopithecine subspecies or allospecies in the tree savannas of western and northern Africa, but no fossils have been found so far. Yet *Homo* must have evolved from some of these peripheral populations. This would explain why *Homo*, a far

TABLE 11.2   Increase of Brain Size in the Hominid Lineage

Species	Body Weight (kg)	Brain Weight (g)
Cercopithecus	4.24	66
Gorilla	126.5	506
Chimpanzee	36.4	410
Australopithecus afarensis	50.6	415
Homo rudolfensis	—	700–900
Homo erectus	58.6	826
Homo sapiens	44.0	1250

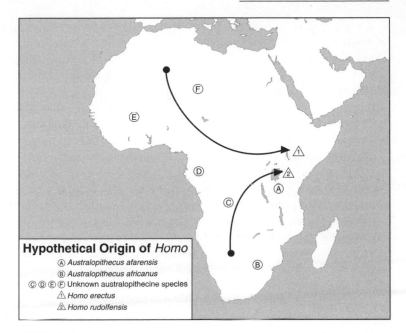

FIGURE 11.3
Hypothetical derivation of *Homo* from australopithecine allospecies.

advanced hominid, appears in eastern Africa so suddenly (Fig. 11.3).
For a different interpretation of the movement of early hominids, see
Strait and Wood (1999). It is based on the assumption that hominids
occurred only in those parts of Africa where fossil hominids have
been found.

A similar history must be inferred for *Homo erectus*, who evidently
originated in Africa about the same time as *H. rudolfensis* but was first
described from Java (1892) and China (1927), because no early fossils
were found in Africa. The earliest representation of the *erectus* lineage
from Africa is *H. ergaster* (1.7 mya), who is perhaps best considered a
subspecies of *H. erectus*. It was this African population that spread
from Africa to Asia presumably sometime between 1.9 and 1.7 mya.

Homo erectus apparently was outstandingly successful. It was the
first hominid to spread out of Africa. Fossils assigned to this species
have been found from eastern Asia (Beijing) and Java, to Georgia

(Caucasus region) (1.7 mya), and to eastern and southern Africa. In addition to being widespread, the species existed without major change for at least 1 million years. The most recent *H. erectus* fossils (ca. 1.0 mya) from Africa indicate a trend toward *H. sapiens.* This fits well with the finding that *H. sapiens* seems to have originated in Africa. *Homo erectus* is characterized by a set of simple stone tools but it evidently succeeded in taming fire. The ability to make use of fire was probably the decisive step in hominization.

The unprecedented rapid increase in brain size took place during the replacement of the tree savanna by the bush savanna. The australopithecines could no longer escape carnivores by climbing trees and so had to depend on their ingenuity. Thus a powerful selection pressure for an increase in brain size developed. This is documented by the brain size of the first fossil *Homo.* The brain of *H. rudolfensis* (1.9 mya) measures 700–900 cc, almost double the size of the *Australopithecus* brain (average of 450 cc). A similar increase of brain size took place in the *H. erectus* lineage, ultimately rising above 1000 cc.

The increase in brain size had a genetic basis and had all sorts of repercussions on the structure of newborn infants and their mothers. A strictly terrestrial mode of life made a positive contribution to the shift. It freed the arms of the mother for new uses other than merely holding on to tree branches, to which the *Australopithecus* mother was restricted. As a result, the australopithecine newborn had to be as advanced as a newborn chimp, who knows how to hold on to its mother. The size of the birth canal through the pelvis allowed the passage of only a small head, and thus the small brain had to be large enough to serve the newborn and the limited demands of an australopithecine.

The most rapid increase in brain size in the whole history of man's ancestors took place when *Homo* originated. *Homo rudolfensis* and *H. erectus* depended for their survival on their ingenuity to cope with their defenseless position in the environment. There must have been a tremendous selection pressure for an increase in brain size, but this increase posed new problems. Infants with larger brains would have larger heads, but as the paleontological record shows, an increase of the size of the birth canal was apparently incompatible with upright posture and bipedal walking. Thus much of the growth of the brain had to be shifted to the postnatal period. In other words, infants had to be born prematurely. Fortunately, the arms of the mothers were no longer needed for climbing and could now be used for the care of their infants. There was now, so to speak, a premium on premature birth. This shift to exclusively terrestrial life must have been a very difficult period in human history. What happened during the transition period involved both infant and mother, for both had to become adapted to the new situation, to the new selection pressures. If the infant had too big a head (brain), it would die owing to labor difficulties. It could survive only if born somewhat prematurely and if rapid growth of the brain was shifted to the postpartum period. At birth, the human newborn is essentially 17 months premature. The mother was also affected in various ways. She had to become bigger to cope with the heavier infant and the long period during which she had to carry it. This led to a striking reduction of the sexual dimorphism in weight.

To put this another way, it is only at the age of 17 months that human infants have acquired the mobility and independence of newborn chimpanzees. But are the somewhat prematurely born human infants fit for survival? For instance, the greatest need of premature human infants is for warmth, so, no doubt, it was for the early *Homo* infants. In response to this selection pressure, they acquired a subcutaneous layer of fat that was a very efficient protection against cold; consequently, they could dispense with a coat of hair. There is no doubt that this shift in the age of birth required a lot of adjustments, particularly of growth rates of both mother and infant. But it permitted the enlarging of brain size in a few million years without an increase in the size of the birth canal. This postponement of brain growth requires that the brain of the human baby almost double in size in the first year after birth.

THE DESCENDANTS OF *HOMO ERECTUS*

As so often occurs in speciational evolution, after an enormous spurt in a short time, *Homo erectus* experienced a period of stasis and, aside from the increase in brain size, not much changed in the evolution from *H. erectus* to *H. sapiens*. The former was the first highly mobile hominid and evolved different geographic races in its wide range from northern China and southeastern Asia to Europe and all of Africa. A remarkably rich fossil record documents a gradual transition from *H. erectus* through *H. heidelbergensis* to Neanderthal. These transitional hominids ranged from England (Swanscomb), Germany (Steinheim), Greece (Petralona), to Java (Ngangdong).

These fossils are best considered "archaic Neanderthal." They changed steadily from being more *erectus*-like to more resembling classic Neanderthal. There is little doubt that, as far as Europe and the Near East are concerned, the western populations of *H. erectus* eventually gave rise to the Neanderthals. But it is still unclear what happened to *H. erectus* in eastern and southern Asia and in Africa.

Neanderthals flourished from about 250,000 to 30,000 years ago. About 100,000 years ago, the range of Neanderthals was overrun by a population wave of *Homo sapiens* that is believed to have come from sub-Saharan Africa, where it had originated about 150,000 to 200,000 years ago. *Homo sapiens* clearly derived from African populations of *H. erectus*. It was presumably isolated from the Asian *H. erectus* for at least half a million years, during which period it acquired the *sapiens* characteristics. A wave of *H. sapiens* eventually broke out of Africa and spread rapidly over the entire world. They reached Australia some 50,000 to 60,000 years ago, eastern Asia 30,000 years ago, and North America reportedly about 12,000 years ago. There is, however, some evidence for an earlier colonization of America, possibly as early as 50,000 years ago.

The hominid chronology in Europe is complex. Fossils of Neanderthal have been found from Turkestan, northern Iran, and Palestine to the entire north coast of the Mediterranean, central Europe, and in western Europe to Spain and Portugal. The study of teeth and cultural remains suggest that Neanderthal was largely carnivorous. No evidence exists that would tell us whether it seriously depleted the megafauna and thereby endangered its own survival. About 35,000 years ago the colonization wave of modern *H. sapiens* reached western Europe, and after several thousand years of coexistence the Neanderthal disappeared. The exact cause or causes of this disappearance (climatic factors, cultural inferiority, genocide by *H. sapiens*) are still controversial. An analysis of mitochondrial DNA showed that the Neanderthal and *H. sapiens* lineage had split around 465,000 years B.C.

The *H. sapiens* invaders of western Europe, called Cro-Magnons, were highly successful but did not change appreciably anatomically, particularly in brain size (1,350 cc), in the nearly 100,000 years of their dominance. They had a highly developed culture, being the creators of the famous paintings in the Lascaux and Chauvet caves.

One can summarize the history of hominid evolution from the ape origin to modern times by emphasizing the drastic reconstruction of man's physique. Most conspicuous is the shift from the semiarboreal mode of living of *Australopithecus* to the strictly terrestrial one of *Homo*. Brain size more than tripled in 4 million years and this facilitated an astounding cultural revolution. The rate of change was not even, but was greatly accelerated in the shift to *Homo*. During the australopithecine phase, no conspicuous change occurred in more than 2 million years. With *Homo*, however, something new appeared, even though there is still some uncertainty about the relationship of *H. habilis*, *H. rudolfensis*, and *H. erectus*. *Homo* was strictly terrestrial and clearly had a larger brain than the apes. But with *H. erectus* another period of stasis was apparently reached, and changes in the 1.5 million years of its existence were relatively minor.

The changes from ape to man in different components of the human phenotype were highly unequal (an example of mosaic evolution). Many of the basic enzymes and other macromolecules, such as hemoglobin, did not change at all. Also, the basic anatomical structure of man is still remarkably similar to that of the chimpanzee, one of the reasons why Linnaeus did not hesitate to place the chimpanzee in the genus *Homo*. Yet there is one structure, the brain, that outpaced all others in its rate of change, beginning about 2.4 million years ago but accelerating during the last half-million years. What is so remarkable about the human brain?

The Brain

The human brain is an unimaginably complex structure. In an adult it contains about thirty billion nerve cells or neurons. The cerebral cortex, which is so highly developed in the human species, contains about ten billion neurons and one million billion connections among them, the so-called synapses. Each neuron has a major stem, its axon, and numerous branchlets, called dendrites, which, in the synapses,

252 WHAT EVOLUTION IS

make contact with other neurons. Much is known about the electro-physiology of the neurons, but very little about their mental functions. The synapses, for instance, apparently play an important role in memory retention, but how they do so is almost entirely unknown.

It has long been appreciated that it is our brain that makes us human. Any other part of our anatomy can be matched or surpassed by a corresponding structure in some other animal. Still, fundamentally, the human brain is very similar to other, far smaller and simpler mammalian brains. The unique character of our brain seems to lie in the existence of many (perhaps as many as forty) different types of neurons, some perhaps specifically human.

What is perhaps most astonishing is the fact that the human brain seems not to have changed one single bit since the first appearance of *Homo sapiens*, some 150,000 years ago. The cultural rise of the human species from primitive hunter-gatherer to agriculture and city civilizations took place without an appreciable increase in brain size. It seems that in an enlarged, more complex society, a bigger brain is no longer rewarded by a reproductive advantage. It certainly shows that there is no teleological trend toward a steady brain increase in the hominid lineage.

It used to be believed that bipedal locomotion and tool use were the most important steps in hominization. The realization of the apelike nature of bipedal *Australopithecus* and the discovery of tool use among chimpanzees (and other animals) have led to an abandonment of this belief. Instead, the rapid growth of the brain seems to have been correlated with two developments in human evolution: the emancipation of hominids from the safety of life in trees and the development of speech, the human system of communication. How did these things come about?

THE UNIQUENESS OF MAN

When it was realized that apes had been man's ancestors, some authors went so far as to state "Man is nothing but an animal." However, this is not at all true. Man is indeed as unique, as different from all other animals, as had been traditionally claimed by theologians and philosophers. This is both our pride and our burden.

I have described the stages by which man became increasingly different from his simian ancestors and must now attempt to describe the characteristics that are uniquely human. Most of them are related to the enormous development of the brain and to the development of extended parental care. In most invertebrates (particularly insects) the parents die before their offspring hatch from the egg. The entire behavioral information available to the newborn is contained in its DNA. What they can subsequently learn during their usually rather short life is quite limited and is not transmitted to their offspring. Only in species with highly developed parental care, as in certain birds and mammals, can the young have an opportunity to add to their genetic information by learning from their parents, as well as from their sibs and occasionally from other members of their social group. Such information can be handed down in these species from generation to generation without being contained in the genetic program. Yet in most animal species the amount of information that can be transferred by such a system of nongenetic information transfer is quite limited. By contrast, in man, the transfer of such cultural information has become a major aspect of life. This capability also favored the development of speech, indeed one might say that it necessitated the origin of language.

Even though we often use the word "language" in connection with the information transmittal systems of animals, such as the "language of bees," actually all of these animal species have merely systems of giving and receiving signals. To be a language, a system of communication must contain syntax and grammar. Psychologists have attempted for half a century to teach language to chimpanzees, but in vain. Chimps seem to lack the neural equipment to adopt syntax. Therefore, they cannot talk about the future or the past. Having invented language, our ancestors were able to develop a rich oral tradition long before the invention of writing and printing. The development of speech, in turn, exerted an enormous selection pressure on an enlargement of the brain, particularly those parts that involved information storage (memory). This enlarged brain made the development of art, literature, mathematics, and science possible.

Thinking and intelligence are widespread among warm-blooded vertebrates (birds and mammals). But human intelligence seems to surpass that of even the most intelligent animals by orders of magnitude. The story that the fossil record tells us of the evolution of the brain is rather surprising. It was originally believed that upright walking had been a major factor in the increase of brain size, by freeing the hands for manipulation. However, the bipedal australopithecines had small brains (mostly below 500 cc), hardly larger than that of the chimpanzee. Then what could have induced the conspicuous increase of brain size in *Homo?* As with so many controversial questions, it is becoming obvious that more than one factor is involved and their major impact may have been at different stages in our history.

The expectation of a smooth continuity of transitional stages in hominization is based on typological thinking. Naturalists had shown, even before Darwin, that higher organisms do not exist as types but as variable populations. They exist as geographically variable species, usually with a central contiguous main body of populations, often surrounded by peripherally isolated incipient species and allospecies. There is much evidence (see Chapter 9) that widespread species undergo relatively little evolutionary change, but that evolutionary novelties occur in the peripheral incipient species. There is every reason to believe that evolution and speciation in the hominids followed the same pattern as in the majority of terrestrial vertebrates.

Peripheral isolates often become so successful that they will overrun the range of the parental species and sometimes even exterminate it. In the fossil record such an event appears as a definite discontinuity, a "saltation" between the parental and the daughter species. In reality, it is only a geographical shift. Let us assume, for instance, that an allospecies of *Australopithecus africanus* from western or northern Africa gradually evolved the characteristics of *Homo*, and then suddenly spread to eastern Africa as *Homo rudolfensis*. There is no conflict between this scenario and the Darwinian explanation, because throughout this process of the geographical speciation of *H. rudolfensis* there was complete populational continuity. The lesson we must learn from this scenario is that one must not look at hominid evolution as a linear typological process in the time dimension, restricted to a single geographical area, but rather as a series of geographical speciation events in a multidimensional sequence. This removes much of the mystery from the process of hominization.

Under severe selection pressure, australopithecine brains grew from less than 500 cc to more than 700 cc. They thus became *Homo*. At this stage in hominid history nothing made a greater contribution to survival than intelligence. *Homo rudolfensis* and *H. erectus* were the first recorded species at this new level of hominization. Curiously, however, brain size after this first spurt by *H. rudolfensis* increased in *H. erectus* only slowly for about one million years, but rose in late *H. erectus* to 800–1,000 cc, finally reaching in *H. sapiens* an average of 1,350 cc. In Neanderthals, who were taller and more robust, brain size reached 1,600 cc, but the relative brain size was a little less than that of *H. sapiens*.

Tool Culture

The different kinds of *Homo* are in part recognized by the tools they manufactured. The earliest stone tools discovered in Africa, referred to as the Oldowan culture, were first ascribed to *Homo habilis*. But now that *H. rudolfensis* is separated from *H. habilis*, this stone culture is credited to *H. rudolfensis*. *Homo erectus* had more elaborate tools, known as the Acheulian culture. It changed remarkably little in the 1.5 million years of the existence of *H. erectus*, but there was some geographic variation. Neanderthals produced more sophisticated tools referred to as Mousterian, and when *H. sapiens* (Cro-Magnon) arrived, its tools, referred to as Aurignac, were considerably superior. It is still unexplained why Aurignacian tools were found in some caves with Neanderthal fossils. Had the Neanderthals traded for them with their Cro-Magnon neighbors?

WHAT IS *HOMO?*
....

The early species *Homo rudolfensis* and *H. erectus* did not reach the brain capacity of the Neanderthals (1,600 cc) or *H. sapiens* (1,350 cc), but the increase from the australopithecine brain of 450 cc to the 700–900 cc of *H. rudolfensis* is almost a doubling of size and a much greater advance than the shift from 900 cc to 1,350 cc, an increase that I do not consider to be of generic value. A genus usually indicates an ecological unit, a noticeable difference in the exploitation of the environment. The designation *Homo* does have such a significance. It designates the emancipation from dependence on trees. Once this independence was achieved, a premium was placed on the enhancement of intelligence, provided the evolutionary unit was small enough to respond to selection. The evolutionary increase of brain size ended when selection for further increase was no longer rewarded by a reproductive advantage.

When our understanding of the mental capacities and emotions of the warm-blooded vertebrates grew in the mid-twentieth century, it led to the successive discovery of astonishing similarities with the human species. However, in the earlier days when most held a belief in the total uniqueness of "Man," the views of anyone calling attention to such similarities were at once labeled as *anthropomorphism*. We are now beginning to realize that such similarities are not surprising, considering our ancestry.

The similarity with warm-blooded members of our vertebrate lineage holds true for most nonphysical human traits. That many kinds of mammals and birds (e.g., corvids, parrots) have a remarkably highly developed intelligence is no longer questioned by psychologists. But it is now realized that many animals also show that they have the emotions of fear, happiness, caution, depression, and almost any other known human emotion. Not every anecdote on such observations in the literature is trustworthy, but there are numerous confirmed cases based on careful observation and testing (Griffin 1981, 1984, 1992; Kaufmann 1981; Masson and McCarthy 1995). Obviously these human characteristics could not have all originated by a big saltation when *Homo sapiens* was born. Naturally, we find the antecedents in many species of animals.

THE EVOLUTION OF HUMAN ETHICS

Few aspects of evolution have been more controversial than the explanation of the origin of human ethics. From 1859 on the objection was raised again and again that altruistic behavior was incompatible with natural selection. It was often asked, Is not selfishness the only behavior that can be rewarded by selection? What is altruism and how can it be defined? Is altruism due to a genetic disposition or is it entirely due to education and learning?

It is perhaps legitimate to admit that real progress in answering these questions was not achieved until analogous behavior was studied in various species of animals. This revealed that one must distinguish between different kinds of altruism and establish different classes of recipients to whom the altruistic behavior is directed.

The traditional definition of altruism is that it consists of an act that is beneficial to the recipient but is performed by the altruist at a cost. This definition excludes all kindness and helpfulness that is performed without noteworthy cost. Yet, in a social group much behavior consists of acts of kindness and thoughtfulness that are performed without any noticeable costs. And it is precisely this kind of behavior that is not only very important for the cohesion of a social group but that also forms a bridge to strictly defined altruism.

Three Kinds of Altruism

When we compare different kinds of altruism, we can distinguish three different classes that differ in the amount and in the evolutionary significance of their altruism.

Altruism for the Benefit of an Individual's Own Offspring. It requires no argument to defend the statement that such altruism would be favored by natural selection. Anything a parent does to enhance the well-being and survival of its offspring favors its own genotype.

Favorite Treatment of Close Relatives (Kin Selection). Most members of a social group are members of an extended family and share part of the same genotype. Any altruism among relatives will be favored by natural selection. This kind of altruism is characteristic for sibs (brothers and sisters) who have known each other from birth and have grown up together. As J. B. S. Haldane was perhaps the first to point out, any support you give to a close relative will add to your own fitness because they have part of your genotype (inclusive fitness selection). The soundness of this conclusion was demonstrated by Hamilton (1964), who applied it to an explanation of the existence of castes in social hymenoptera. The question whether more distant relatives are also favored is controversial.

Altruism Among Members of the Same Social Group.   Social groups usually consist not only of members of an extended family but also of "immigrants," outsiders who had transferred from another group in search of attachment. In a social group, members seem to realize that additional workers or potential breeders would sometimes strengthen the group and are therefore usually somewhat tolerant toward such newcomers. Indeed, it is likely that the development of friendly and cooperative feelings among all members of the same social group is favored by natural selection. It is not quite certain how much greater in a social group the altruism is between related individuals (kin selection) and other members of the group.

Reciprocal Helpfulness.   The cohesion of a social group is enhanced by reciprocal helpfulness. It is often observed among social animals that an individual will help another one in the expectation that at some future occasion the recipient will return this favor. This behavior is usually referred to as reciprocal altruism, but owing to the expected reciprocity, the motivation of such helpfulness is evidently selfish. Such mutual helpfulness is found not only among members of the same social group, but sometimes also among members of different groups, indeed occasionally also between members of different species. The "cleaner fishes" that free large predatory fishes of external parasites (admittedly in exchange for food and protection) illustrate such interspecific helpfulness. One might even go so far as to include the whole range of symbiotic interactions in this category.

Behavior Toward Outsiders.   The same kinds of altruism that are extended to other members of a social group are rarely offered to outsiders. Different social groups usually compete with each other and not infrequently fight each other There is little doubt that hominid history is a history of genocide. Indeed, the same can apparently be said about chimpanzees. How then could the propensity for altruistic behavior of members of a social group be redirected to include among the recipients of their altruism individuals that are not members of the group? How can such altruism to outsiders be created? Evidently genuine ethics can be developed only by adding such global altruism to the "selfish" altruism of the social group.

How could such altruism toward outsiders have become established in the human species? Could natural selection be invoked? This has often been tried, but not very successfully. It is difficult to construct a scenario in which benevolent behavior toward competitors and enemies could be rewarded by natural selection. It is interesting in this connection to read the Old Testament and see how consistently a difference is made between behavior toward one's own group and behavior to any outsiders. This is in total contrast to the ethics promoted in the New Testament. Jesus's parable of the altruism of the Good Samaritan was a striking departure from custom. Altruism toward strangers is a behavior not supported by natural selection.

The propensity for altruistic behavior toward other insiders of the social group is an all-important component in the evolution of genuine ethics. But it requires a cultural factor, the preaching of a religious leader or a philosopher, to be implemented. It is not automatically produced by evolution. Genuine ethics is the result of the thought of cultural leaders. We are not born with a feeling of altruism toward outsiders, but acquire it through cultural learning. It requires the redirecting of our inborn altruistic tendencies toward a new target: outsiders.

There is great variation in the altruistic propensity of different individuals. Occasionally, we encounter a person with an exceptional capacity for human kindness, altruism, generosity, and cooperativeness. And the families in which these individuals occur always insist that these individuals had been that way from infancy on. But we also know that there is an opposite extreme, sociopathy. Many criminals have such a pathological propensity, and all efforts of education are usually rather unsuccessful with such individuals. Most individuals, however, lie somewhere in between these two extremes. They acquire true ethics (including to outsiders) by learning. The low rate of criminality in Utah, a state in which Mormon ethical principles are widely adopted, documents this effect of learning.

The promoters of ethical principles for mankind had a difficult uphill struggle, for the inborn suspicion and hostility against aliens (outsiders) are difficult to overcome. But there were also factors that aided in the adoption of ethics. Reciprocal helpfulness worked as successfully with outsiders as with group members. Yet far more important was the diversity within human populations. Every population contains individuals with a particularly friendly disposition, and they help in the making of bridges between groups and populations. This diversity, and the recognition of it, is helping to refute the rigid typological interpretation of terms such as race.

The discrimination against outsiders, which is perhaps the major reason for the resistance to a worldwide acceptance of a broadly conceived human ethics, is gradually being overcome by some basic social principles, such as equality, democracy, tolerance, and human rights. Moral education has been practiced very successfully by several of the world's great religions. And where these religions have failed, as in the prevention of the two appalling world wars, we can hope that the world has learned from past mistakes. And, let us appreciate it, the cultures of the Christian world do have ethical principles that are, on the whole, perfectly sound, even though we have failed so often to follow them.

MAN AND HIS ENVIRONMENT

Our superb brain has enabled us to create one invention after the other by which to become increasingly independent of the environment. No other animal was ever able to exist successfully on all continents and in all climates. No other animal has ever achieved the same relative dominance over nature. But in the last 50 years it has become evident that we are still thoroughly dependent on the natural world and that our efforts at dominating nature carry a high price. Some of these mounting costs include the overexploitation of nonrenewable resources and the continuing destruction of the sources of renewable resources. They consist of air and water pollution, the accelerating destruction of natural environments and of the fruits of evolution—the diversity of plant and animal life—and the development of such appalling social conditions as slums, poverty, and shanty towns (Ehrlich 2000).

THE FUTURE OF MANKIND
....

Two questions are frequently asked about the future of mankind. The first one is, What is the probability that the human species will break up into several species? The answer is clear: none at all. Humans occupy all the conceivable niches from the Arctic to the tropics that a humanlike animal might occupy. Furthermore, there is no geographic isolation between any of the human populations. Whenever geographically isolated human races developed in the last 100,000 years, they interbred readily with other races as soon as contact was reestablished. Today there is far too much contact among all human populations for any kind of effective long-term isolation that might lead to speciation.

The second question is, Could the now existing human species evolve as a whole into a "better" new species? Could Man become superman? Here again, one cannot be hopeful. To be sure, there is abundant genetic variation within the human genotype to serve as material for appropriate selection, but modern conditions are very different from the time when some populations of *Homo erectus* evolved into *Homo sapiens*. At that time, our species consisted of small troops, in each of which there was strong natural selection with a premium on those characteristics that eventually resulted in *Homo sapiens*. Furthermore, as in most social animals, there was undoubtedly strong group selection.

Modern humans, by contrast, constitute a mass society and there is no indication of any natural selection for superior genotypes that would permit the rise of the human species above its present capacities. With selection for improvement no longer being exercised, there is no chance for the evolution of a superior human species. Indeed, some students of this problem fear that a deterioration of our species is inevitable under the conditions of a mass society. However, genetic deterioration is not an immediate danger, considering the high variability of the human gene pool.

Are There Human Races?

When one compares an Inuit with an African Bushman, or a Nilotic Negro, an Australian aborigine, a Chinese, or a blond, blue-eyed northern European, one cannot escape recognizing the so-called racial differences. But does this not conflict with our fervent belief in human equality? No, it does not, provided we define both equality and race properly.

Equality is civic equality. It means equality before the law and it means equal opportunity. But it does not mean total identity, because we now know that every one of the 6 billion human individuals is genetically unique. Not every human has the mathematical ability of an Einstein or the speed of an Olympic sprinter, nor the imagination of a good novelist or the aesthetic sense of an outstanding painter. Every parent knows that each of his or her children is uniquely different. The time has come that we must honestly face and admit these differences. What is important is to realize that these differences also exist within all of the human races.

The major reason for the existence of a race problem is that so many people have a faulty understanding of race. These people are typologists, and for them every member of a race has all the actual and imaginary characteristics of that race. To translate this bias into an absurd example, they would assume that every African-American can run the 100-meter dash faster than any European-American. Yet, if in a racially mixed class in a school the students were seated according to their performance in various mental, physical, manual, and artistic challenges, each ranking would be different and each "race" would be distributed through a greater part of the ranking. In other words, by rejecting the typological approach, which considers the members of each race as a type, and replacing it by the populational approach in which each individual is considered on the basis of his or her particular abilities, a truer understanding of reality can be achieved that avoids any typological ranking and any discrimination based on such ranking.

Are Humans Alone?

The question is often asked, Are we the only intelligent beings in this vast universe? If we want to find an answer to this question, we must break the question down into a number of components. Where could life exist? Only on planets, because suns are far too hot. Surely, many stars must have planets, but it is only in the last 20 years that planets beyond the solar system were discovered. But so far all of them turned out to be quite unsuitable for the origin and maintenance of life. The set of conditions found on the Earth (and perhaps at one time also on Mars and Venus) that makes life possible is seemingly quite exceptional. Even so, considering the high number of planets, some of them probably do have conditions suitable for the origin of life.

But what is the chance of life originating on a suitable planet? Apparently quite high. Many of the kinds of molecules needed for the origin of life are widely distributed in the universe, including purines, pyrimidines, and amino acids. It has even been shown experimentally in the laboratory that under certain anoxic atmospheric conditions rather complex organic molecules may be produced spontaneously from simpler molecules. Therefore, it is quite conceivable that some primitive forms of life have repeatedly originated on other planets. If such evolution was successful, it might ultimately result in bacteria-alike organisms.

Alas, the rutted road from bacteria to humans is long and difficult. Following the origin of life on Earth there were nothing but prokaryotes for the next billion years, and highly intelligent life originated only about 300,000 years ago, in a single one of the more than one billion species that had arisen on Earth. These are indeed long odds.

Even if something parallel to the origin of human intelligence should indeed have happened somewhere in the infinite universe, the chance that we would be able to communicate with it must be considered as zero. Yes, for all practical purposes, man is alone.

END
••••••

Evolution is often considered as something unexpected. Wouldn't it be more natural, some antievolutionists ask, if everything would always stay the same? Perhaps this was a valid question before we understood genetics, but it is no longer. In fact, the way organisms are structured, evolution is inevitable. Each organism, even the simplest bacterium, has a genome, consisting of thousands to many millions of base pairs. Observation has established that each base pair is subject to occasional mutation. Different populations have different mutations, and if they are isolated from each other, these populations inevitably become more different from each other from generation to generation. Even this simplest of all possible scenarios represents evolution. If one adds further biological processes, such as recombination and selection, the rate of evolution accelerates exponentially. Therefore, the mere fact of the existence of genetic programs makes the assumption of a stationary world impossible. Evolution is thus a plain fact, not a conjecture or assumption.

It is very questionable whether the term "evolutionary theory" should be used any longer. That evolution has occurred and takes place all the time is a fact so overwhelmingly established that it has become irrational to call it a theory. To be sure, there are particular evolutionary theories such as those of common descent, origin of life, gradualism, speciation, and natural selection, but scientific arguments about conflicting theories concerning these topics do not in any way affect the basic conclusion that evolution as such is a fact. It has taken place ever since the origin of life.

··

CHAPTER 12

THE FRONTIERS OF EVOLUTIONARY BIOLOGY

Everyone knows that our understanding of this world is still incomplete, in spite of the magnificent advances of science. We must ask ourselves, therefore, to what extent this is also true of evolutionary biology.

Here it must be emphasized that the development of molecular biology has resulted in an enormous increase of interest in and understanding of evolution. Easily one-third, if not more, of all papers now published in molecular biology deal with evolutionary questions. Molecular techniques allow us to solve numerous problems that had previously been inaccessible. This is particularly true for phylogenetic problems, for issues in the chronology of evolution, and for the role of development in evolution.

When we look back over the controversies of the last 140 years, what is most impressive is the robustness of the original Darwinian paradigm. The three major theories that competed with it—transmutationism, Lamarckism, and orthogenesis—were decisively refuted by about 1940, and no viable alternative to Darwinism has been proposed in the last 60 years. But this does not mean that we possess a full understanding of all aspects of evolution. I shall now try to enumerate a number of evolutionary phenomena in need of further research and explanation.

To begin with, we still have only a very incomplete knowledge of biodiversity. Although nearly two million animals have already been described, estimates of the number of still undescribed species go as high as 30 million. Fungi, lower plants, protists, and prokaryotes are even more poorly known. The phylogenetic relationship of most of these taxa are only poorly understood or entirely unknown, although molecular methods now make daily new contributions to this understanding. The fossil record of past evolution is still woefully inadequate, as illustrated by the hominid fossil record. Almost every month some new fossil is found somewhere in the world that solves an old problem, or poses a new one. And the ups and downs of the former biota raise innumerable questions about the causes of mass extinction and the varying fates of different phyletic lineages and higher taxa. Even at this rather descriptive level, our ignorance is still enormous. But there are also many uncertainties about aspects of evolutionary theory.

Even though there is no doubt as to the prevalence of geographic (allopatric) speciation and (in plants) polyploidy as the prevailing forms of speciation, we are still uncertain about the frequency of other forms of speciation, for instance, sympatric speciation. The contribution of various factors to the extraordinary rapidity (less than 10,000 or even 1,000 years) of speciation in certain groups of fishes is still not understood.

The astonishing slowdown or stasis of certain evolutionary lineages ("living fossils") is also rather puzzling, considering that all the other members of their biota evolved at normal rates. The opposite extreme, the rapidity with which certain genotypes were restructured in founder populations, is likewise puzzling.

All of these puzzling problems ultimately seem to be due to the structure of the genotype. Molecular biology has discovered that there is a variety of kinds of genes, some in charge of the production of certain materials (enzymes), others involved in the regulation of the activity of other genes. Most genes apparently are not continuously active but only in certain cells (tissues) and at certain times in the life cycle. Other genes seem to be neutral, while an amazingly large proportion of DNA seems to be totally inactive. The genes of the genotype, therefore, form a complex system of interactions. Owing to these multiple interactions among all the composing genes, such a system is highly constrained. It can respond to some influences or environmental pressures, though most would lead to unbalances and will be selected against.

There are suggestions that genotypes were less tightly constrained at the beginning of the existence of the Metazoa so that for 200–300 million years in the late Precambrian or early Cambrian no fewer than 70 or 80 new structural types evolved. Only about 35 are now left, none of which has changed drastically (in the basics of their body plan) in the 500 million years since the Cambrian. How can we explain such a seemingly drastic change in evolutionary rate? Within these surviving structural types, however, there have been remarkable radiations, such as the insects and the vertebrates.

THE USEFULNESS OF EVOLUTIONARY THOUGHT
....

Evolutionary thought, and in particular an understanding of the new concepts developed in evolutionary biology, such as population, biological species, coevolution, adaptation, and competition, is indispensable for most human activities. We apply evolutionary thinking and evolutionary models to cope with antibiotic resistance by pathogens, pesticide resistance by crop pests, the control of disease vectors (e.g., malaria mosquitoes), human epidemics, the production of new crop plants by evolutionary genetics, and many more challenges (Futuyma 1998: 6–9).

The principal reason why scientists study evolution is to further our understanding of this phenomenon that affects every aspect of the living world. But evolutionary studies have also made many important contributions to human welfare. Evolutionary thinking has enormously enriched almost all other branches of biology. For instance, more than a third of all current publications in molecular biology show how the nature and history of important biological molecules are illuminated by an evolutionary approach. Developmental biology has been completely revitalized by the study of evolutionary questions and the establishment of the different categories of genes and their elaboration in the course of phylogeny. The evolutionary approach has also given us a wonderful insight into the history of mankind. And nothing has contributed more to our understanding of such human characteristics as mind, consciousness, altruism, character traits, and emotion than comparative studies of the behavior of animals.

It must never be forgotten that the genotype is a harmonious, interacting system that is exposed to natural selection as a whole. Whenever it is inferior in competition with some other genotype it will be selected against, a process that may lead to the extinction of the inferior species.

Biology also tries to explain three other complex systems: the developmental system, the neurosystem, and the ecosystem. Three major biological disciplines are occupied with this task. The study of the developmental system is the task of developmental biology; that of the neurosystem (central nervous system) is the task of neurobiology; and that of the ecosystem is the purview of ecology. However, in all three cases, it is the structure of the genotype that is ultimately responsible for how organisms can meet the challenges of these three systems. Our knowledge of the underlying building blocks of all three systems is already well advanced. Where we are deficient in explanations is the control of the interactions of the components of these systems. No doubt evolutionary biology will make major contributions toward this end.

··

WHAT CRITICISMS HAVE BEEN MADE OF EVOLUTIONARY THEORY?

The story of evolution as it was worked out during the past fifty years continues to be attacked and criticized. The critics either hold an entirely different ideology, as do the creationists, or they simply misunderstand the Darwinian paradigm. An author who says, "I cannot believe that the eye evolved through a series of accidents," documents that he or she simply does not understand the two-step nature of natural selection. A typologist, not used to population thinking, will indeed have major difficulties appreciating the amount of genetic variability available to selection in natural populations.

All theories of Darwinism are subject to rejection if they are falsified. They are not unalterable like the revealed dogmas of religions. The history of evolutionary biology reports numerous cases of evolutionary theories that were eventually rejected. The belief that a gene can be the direct object of selection is one such refuted theory. An inheritance of acquired characters is another one.

In the preceding chapters I endeavored to present the story of the phenomena and processes of evolution as seen by contemporary evolutionists. These conclusions are not accepted by everybody, and it might be worthwhile to give a short summary of some of the criticisms and the responses to them by the evolutionists. I also discuss some biological phenomena considered by some authors to be in conflict with Darwinism.

Creationists

The claims of the creationists have been refuted so frequently and so thoroughly that there is no need to cover this subject once more. I refer to the publications by Alters, Eldredge, Futuyma, Kitcher, Montagu, Newell, Peacocke, Ruse, and Young listed in the bibliography (see Box 1).

Punctuated Equilibria

The claim has been made by some authors (Gould 1977) that the occurrence of punctuated equilibria is in conflict with gradual Darwinian evolution. This is not correct. Even punctuated equilibria, which, at first sight, seem to support saltationism and discontinuity, are in fact strictly populational phenomena, and therefore gradual (Mayr 1963). They are in no respect whatsoever in conflict with the conclusions of the evolutionary synthesis (see Chapter 10).

Neutral Evolution

It was claimed by Kimura (1983) and others that neutral evolution is in conflict with Darwinism. This is not correct, since the assumption in the theory of neutral evolution is that the gene is the object of selection, rather than the individual. However, in reality, it is the individual as a whole that is the target. Under these circumstances, there is no conflict with Darwinism when in the selection of certain favorite individuals some neutral gene replacements may be transmitted to the next generation as incidental components of the favored genotype (see Chapter 10).

Morphogenesis

It is claimed by some authors that the phenomena of morphogenesis, and in particular the processes of development, are in conflict with Darwinism. Even though many of the causal phenomena of development are still insufficiently understood, what is understood is entirely compatible with a Darwinian explanation. It seems that some of those who raise these criticisms assume that only the adult phenotype, the last stage of development, is exposed to selection. In reality, every stage of a developing organism, from the fertilized egg (zygote) on to old age, is constantly subjected to selection. However, the fate of postreproductive individuals is of no relevance to evolution (see Chapter 6).

Causes for Misunderstandings

There are a number of reasons why the evolutionary process is so often misunderstood. Let us look at some of these reasons.

Multiple simultaneous causations. Authors frequently look at only one of the causations of a particular evolutionary phenomenon: either the proximate or the evolutionary causation. This may lead to wrong conclusions because any evolutionary phenomenon is the result of the simultaneous occurrence of both proximate and ultimate causations. Multiple causations are involved in all selection processes because chance phenomena occur simultaneously with selection. To give an example, speciation is never merely a matter of genes or chromosomes, but also of the nature and geography of the populations in which the genetic changes occur. Geography and the genetic changes in populations affect the speciation process simultaneously.

Pluralistic solutions. Almost all evolutionary challenges have multiple solutions. During speciation, for instance, premating isolating mechanisms originate first in some groups of organisms, and in others postmating mechanisms originate first. Geographic races are sometimes phenotypically as distinct as true species without being reproductively isolated; on the other hand, phenotypically indistinguishable species (sibling species) may be fully isolated genetically. Polyploidy and asexual reproduction are important in some groups of organisms, but totally absent in others. Chromosomal reconstruction seems to be an important component of speciation in some groups of organisms, but does not occur in others. Some groups speciate profusely, whereas in others speciation seems to be a rare event. Gene flow is rampant is some species, but drastically reduced in others. One phyletic lineage may evolve very rapidly, while geographically isolated species may remain in complete stasis for many millions of years. In short, there are multiple possible solutions to most evolutionary challenges, even though all of them are compatible with the Darwinian paradigm. The lesson one must learn from this pluralism is that sweeping generalizations are rarely correct in evolutionary biology. Even when something occurs "usually," this does not mean that it must always occur (see Chapter 10).

Mosaic evolution. I have repeatedly called attention to the highly variable rates of evolution. This is true not only for sister lineages, but also for components of a single genotype. As an example, I discussed the divergence between chimpanzee and man since their descent from a common ancestor. In this case some of the protein genes have not changed at all, while those that contribute in the human lineage to the development of the central nervous system have undergone extremely rapid evolution. Why some lineages seem to be able to enter a stage of complete stasis ("living fossils"), lasting for many millions of years, is still not understood (see Chapter 10).

The Findings of Molecular Biology

It is sometimes claimed that the findings of molecular biology necessitate a complete revision of the Darwinian theory. This is not the case. All the findings of molecular biology relevant to evolution deal with the nature and the origin of genetic variation. Even though this includes some unexpected phenomena, such as transposons (genes that can "jump" from one chromosome or position to another), they merely affect the nature and the amount of the available variation, and all of this variation is ultimately exposed to natural selection, and thus is part of the Darwinian process. The molecular discoveries of the greatest evolutionary importance are the following:

1. The genetic program (DNA) does not by itself provide the building material of a new organism, but is only a blueprint (information) for making the proteins of the phenotype.
2. The pathway from nucleic acids to proteins is a one-way street. Proteins and information contained in them cannot be translated back into nucleic acids.
3. Not only the genetic code, but in fact most of the basic molecular cellular mechanisms are the same in all organisms from the most primitive prokaryotes to humans (see Chapter 5).

Unanswered Questions

Darwinian evolutionists have every reason to be proud of the paradigm of evolutionary biology that they have constructed. Every attempt in the last 50 years to refute one or the other assumption of Darwinism has been invalidated. Furthermore, no competing evolutionary theory has been proposed, certainly none that was in any way successful. Does this mean that we now fully understand the evolutionary process in all of its details? The answer to this question is a qualified "No."

In particular, there is one problem that is not yet entirely solved. When we look at what happens to the genotype during evolutionary change, particularly relating to such extreme phenomena as highly rapid evolution and complete stasis, we must admit that we do not fully understand them. The reason for this is that evolution is not a matter of changes in single genes; evolution consists of the change of entire genotypes. It was realized rather early in the history of genetics that most genes are pleiotropic, that is, a single gene may have simultaneous effects on several aspects of the phenotype. Likewise, it was found that most components of the phenotype are polygenically determined, that is, are affected by multiple genes. Such frequent, in fact universal, interactions among genes are of decisive importance for the fitness of individuals and for the effects of selection. Yet, they are singularly difficult to analyze. Most population genetics still focuses on additive gene effects and on the analysis of single gene loci. This is why the study of phenomena such as evolutionary stasis and the constancy of body plans is so refractory to analysis. There are many indications that separate domains exist within a genotype and that certain gene complexes have an internal cohesion that resists breakage by recombination. Up to now, however, these are only ideas; their genetic analysis still lies in the future. The structure of the genotype is perhaps the most challenging remaining problem of evolutionary biology.

SHORT ANSWERS TO FREQUENTLY ASKED QUESTIONS ABOUT EVOLUTION

1. Is evolution a fact?
2. Does any process in evolution require a teleological explanation?
3. What is the Darwinian theory?
4. How do the "facts" of evolution differ from those of physics?
5. How can evolutionary theories be established?
6. Is Darwinism an unalterable dogma?
7. Why is evolution unpredictable?
8. What did the evolutionary synthesis achieve?
9. Have the molecular discoveries required a change of the Darwinian paradigm?
10. Are the words "evolution" and "phylogeny" synonyms?
11. Is evolution progressive?
12. How can long-lasting stasis be explained?
13. How can we explain the two great puzzles in the phylogeny of animals?
14. Is the Gaia hypothesis incompatible with Darwinism?
15. What is the role of mutation in evolution?
16. Is species selection a valid concept?
17. Can the statement that the individual is ordinarily the target of selection also be applied to asexual organisms?
18. What is the object of natural selection?
19. At what stages of development is the individual an object of selection?
20. Is the term "struggle for existence" to be interpreted literally?
21. Is selection a force or a pressure?

22. Where does chance (stochastic processes) enter the process of selection?
23. Does selection produce perfection?
24. How did human consciousness evolve?

The story of evolution is so diversified that it poses innumerable questions to whoever first encounters evolutionary problems. Even though I have attempted, in the first twelve chapters, to answer these questions in considerable detail, I now try to provide a concise answer to the most frequently posed of these questions.

1. Is evolution a fact?
Evolution is not merely an idea, a theory, or a concept, but is the name of a process in nature, the occurrence of which can be documented by mountains of evidence that nobody has been able to refute. Some of this evidence was summarized in Chapters 1–3. It is now actually misleading to refer to evolution as a theory, considering the massive evidence that has been discovered over the last 140 years documenting its existence. Evolution is no longer a theory, it is simply a fact.

2. Does any process in evolution require a teleological explanation?
The answer is an emphatic "No." In earlier periods many authors thought that a perfection-giving process was involved in evolution. Before the discovery of the principle of natural selection, one could not imagine any other principle than teleology that would lead to such seemingly perfect organs as the eye, annual migrations, certain kinds of disease resistance, and other properties of organisms. However, orthogenesis and other teleological explanations of evolution have now been thoroughly refuted, and it has been shown that indeed natural selection is capable of producing all the adaptations that were formerly attributed to orthogenesis (see Chapters 6 and 7).

3. What is the Darwinian theory?
This is the wrong question. In *On the Origin of Species* and in his later publications, Darwin advanced numerous theories, among which five are most important (see Chapter 4). Two of them, evolution as such and the theory of common descent, were accepted by biologists within a few years of the publication of the *Origin* in 1859 (see Box 5.1). This was the first Darwinian revolution. The other three theories, gradualism, speciation, and natural selection, were widely accepted only much later, during the time of the evolutionary synthesis in the 1940s. This was the second Darwinian revolution.

4. Are not the "facts" of evolutionary biology something very different from the facts of astronomy, which show that the Earth circles the sun rather than the reverse?

Yes, up to a point. The movement of planets can be observed directly. By contrast, evolution is a historical process. Past stages cannot be observed directly, but must be inferred from the context. Yet these inferences have enormous certainty because (1) the answers can very often be predicted and the actual findings then confirm them, (2) the answers can be confirmed by several different lines of evidence, and (3) in most cases no rational alternative explanation can be found.

If, for instance, in a chronological series of geological strata a series of fossil therapsid reptiles is found that become more and more similar to mammals in successively younger strata, finally producing species about which specialists argue whether they are still reptiles or already mammals, then I do not know of any other reasonable explanation than that mammals evolved from therapsid ancestors. Actually, there are thousands of such series in the fossil record, even though admittedly there are occasional breaks in most of these series, owing to breaks in the fossil-bearing stratigraphy.

Frankly, I cannot see why such an overwhelming number of well-substantiated inferences is not scientifically as convincing as direct observations. Many theories in other historical sciences, such as geology and cosmology, are also based on inferences. The endeavor of certain philosophers to construct a fundamental difference between the two kinds of evidence strikes me as misleading.

5. How can we establish theories concerning the causes of historical evolutionary processes when the most common method of science, the experiment, cannot be employed?

It is obvious, for example, that we cannot experiment with the extinction of the dinosaurs. Instead, one applies the method of "historical narratives" to explain historical (including evolutionary) processes. That is, one proposes an assumed historical scenario as a possible explanation and tests it thoroughly for the probability of its correctness. In the case of the extinction of the dinosaurs, a number of possible scenarios were tested (such as a devastating virus epidemic or a climatic disaster) but rejected because they were found to be in conflict with the evidence. Finally, the Alvarez extinction theory (caused by an asteroid impact) was so convincingly supported by the existing evidence and by all subsequent research that it is now universally accepted (see Chapter 10).

6. Is Darwinism an unalterable dogma?

All theories of science, including Darwinism, are vulnerable to rejection if they are falsified. They are not unalterable, in contrast to revealed dogmas of religions. There are numerous cases in the evolutionary literature of provisional evolutionary theories that were eventually rejected. The belief that a gene can be the direct object of selection is one such refuted theory. The formerly widely adopted theories of transmutationism and transformationism were also rejected.

7. Why is evolution unpredictable?

Evolution is subject to a large number of interactions. Different genotypes within a single population may respond differently to the same change of the environment. The changes of the environment, likewise, are unpredictable, particularly the arrival at a locality of new predators and competitors. Finally, there are occasionally very drastic changes in the global environment, resulting in so-called mass extinctions. In such mass events, chance may play a large role in survival. Owing to the unpredictability of all of these situations, the nature of the evolutionary change by which a population will respond is necessarily also unpredictable. Nevertheless, a knowledge of the potential of a genotype and of the nature of constraints permits in most cases a reasonably accurate prediction.

8. What did the evolutionary synthesis achieve?

Three accomplishments of the synthesis are particularly important. First, it effected the universal rejection of the three evolutionary theories competing with Darwinism, orthogenesis (finalism), transmutationism (based on saltations), and inheritance of acquired characters; second, it produced a synthesis between the thinking of the students of adaptation (anagenesis) and those of organic diversity (cladogenesis); and third, it confirmed the original Darwinian paradigm of variation and selection while refuting all criticism of it.

9. Have the molecular discoveries required a change of the Darwinian paradigm?

Molecular biology has made great contributions to our understanding of the evolutionary process. However, the basic Darwinian concepts of variation and selection were not affected in any way. Not even the replacement of proteins by nucleic acids as the carriers of the genetic information required a change in the evolutionary theory. Indeed, an understanding of the nature of genetic variation has contributed greatly to strengthening Darwinism. For instance, it confirmed the finding of the geneticists that an inheritance of acquired characters is impossible. Also the use of molecular evidence when added to the morphological evidence has led to the solution of many phylogenetic puzzles.

10. Are the words "evolution" and "phylogeny" synonyms?

No, evolution is a much broader concept. Phylogeny refers only to one of many evolutionary phenomena, the pattern of common descent. However, properly considered, phylogeny means not only the pattern of branching points, but also the changes between these nodes.

11. Is evolution progressive?

Are phylogenetically later organisms "higher" than their ancestors? Yes, they are higher on the phylogenetic tree. But is it true that they are "better" than their ancestors? Those who make this claim list a number of characteristics of "higher" organisms, purporting to demonstrate advance, such as division of labor among their organs, differentiation, greater complexity, better utilization of the resources of the environment, and in general better adaptation. But are these so-called measures of "progress" truly valid evidence for an advance?

It seems that those who deny any signs of evolutionary progress in the advance from bacteria to higher organisms give a teleological or deterministic aspect to the idea of progress. Indeed, evolution seems highly progressive when we look at the lineage leading from bacteria to cellular protists, higher plants and animals, primates, and man. However, the earliest of these organisms, the bacteria, are just about the most successful of all organisms, with a total biomass that may well exceed that of all other organisms combined. Furthermore, among the higher organisms there are lineages such as parasites, cave animals, subterranean animals, and other specialists that show many retrogressive and simplifying trends. They may be higher on the phylogenetic tree, but they lack the characteristics always listed as evidence for evolutionary progress. What cannot be denied, however, is that in every generation of the evolutionary process, a surviving individual is on the average better adapted than the average of the nonsurvivors. To that extent, evolution clearly is progressive. Also, throughout evolutionary history innovations were introduced that made functional processes more efficient.

12. How can long-lasting stasis be explained?

Once a species has acquired effective isolating mechanisms, it may not materially change for millions of years. Indeed the so-called living fossils have hardly changed for hundreds of millions of years. How can this be explained? It has been argued that this stasis was due to the operation of normalizing selection, which culls all the deviations from the optimal genotype. However, normalizing selection is equally active in rapidly evolving lineages. Stasis apparently indicates the possession of a genotype that is able to adjust to all changes of the environment without the need for changing its basic phenotype. To explain how this is done is the task of developmental genetics.

13. How can we explain the two great puzzles in the phylogeny of animals?

The first puzzle is the sudden appearance of 60 to 80 different structural types (body plans) of animals in the early Cambrian, and the second puzzle is why no major new types originated in the 500 million years since the Cambrian.

It is now clear that the seemingly sudden origin (within 10–20 million years) of so many animal types in the early Cambrian (beginning 544 million years ago) is an artifact of preservation. By use of the molecular clock, the origin of the animal types can be placed at about 670 million years ago, but the animals living between 670 and 544 million years ago are not preserved as fossils because they were very small and without skeletons.

The reason why no major new types originated in the ensuing 500 million years is more complex and only partly understood. However, molecular genetics has led to an explanatory suggestion. Development is tightly controlled in the now living organisms by very precise "working teams" of regulatory genes. In the Precambrian, there were apparently only a few such genes, which did not control development as tightly as later on. This allowed a frequent occurrence of rapid major restructuring of the structural types. By the end of the Cambrian, the dominance of these regulatory genes had been fully established and the origin of completely new structural types had become difficult, if not impossible. One must always remember that the changes prior to the Cambrian did not occur suddenly, but over a period of several hundred million years, even though not documented in the fossil record.

14. Is the Gaia hypothesis incompatible with Darwinism?

Even though most Darwinians do not accept the Gaia hypothesis, the most prominent adherents of the Gaia hypothesis, for instance, Lynn Margulis, completely accept Darwinism. There is no conflict.

15. What is the role of mutation in evolution?

Mutation is the principal source of new genetic variation in a population. Most mutations are due to errors in the replication process during meiosis that are not corrected by any repair mechanism. There is no mutation pressure. Most of the variation of genotypes available for selection in a population is the result of recombination, not of new mutations.

16. Is species selection a valid concept?
Darwin already pointed out that the introduction of English plants and animals on New Zealand often resulted in the extinction of native species. Indeed, it has frequently been observed in other parts of the world that the success of one species may result in the downfall of another species. Authors have spoken of species selection, but this is a misleading term. Actually, selection acts on the individuals of the two species as if they were members of a single population. Therefore, the "struggle for existence" is between the individuals of the two species, but the individuals of one of them are in the long run more successful than those of the other species. Thus it is a typical case of Darwinian selection of individuals. The species as a whole is never the target of selection. However, one can admit that the differential success of entire species is superimposed on this individual selection. Misunderstanding can be avoided if one speaks of species turnover or species replacement, instead of species selection.

17. Can the statement that the individual is ordinarily the target of selection also be applied to asexual organisms?
The individual in an asexually reproducing organism is the entire clone, that is, the totality of genetically identical individuals. Such an individual is replaced by selection at the moment at which the last member of the clone dies. Such an elimination is in principle the same as the elimination of an individual by natural selection in sexual organisms.

18. What is the object of natural selection?
Why has there been so much controversy about the object of selection? At the time of the evolutionary synthesis, the geneticists believed that it was the gene, whereas the naturalists believed that it was the individual, as Darwin had always believed. Forty years of analysis have finally made it quite clear that the gene as such could never be the direct target of selection. However, in addition to the individual, a group can also be the target of selection if it is a social group and cooperation within this group enhances its survival. Finally, gametes are also directly exposed to selection and different gametes produced by the same individual may differ in their ability to achieve fertilization.

19. At what stages of development is the individual an object of selection?

From the stage of the zygote on. Some evolutionists have neglected to take embryonic or larval life into consideration. These are often subjected to more selection pressure than the adults. However, the evolutionary effectiveness of selection ends with the end of the reproductive life. In the human species, for instance, diseases that manifest themselves only in postreproductive life are virtually unaffected by selection. Yet they may reduce the contribution to kin selection made by healthy grandparents (in social organisms).

20. Is the term "struggle for existence" to be interpreted literally?
Definitely not! As Darwin already emphasized, the term is to be interpreted metaphorically. Plants at the edge of the desert may struggle for existence with each other, as few will survive while most of them will succumb to the desert conditions. However, a literal struggle is quite rare. It does occur in polygynous species of animals in which males fight with each other in territorial encounters, and it also occurs in struggle for space among marine benthic organisms, and in similar situations. It is most obvious whenever competition for space is involved. In social organisms, low-ranking individuals may struggle for resources with high-ranking individuals.

21. Is selection a force or a pressure?
In evolutionary discussions, it is often stated that "selection pressure" resulted in the success or elimination of certain characteristics. Evolutionists here have used terminology from the physical sciences. What is meant, of course, is simply that a consistent lack of success of certain phenotypes and their elimination from the population result in the observed changes in a population. It must be remembered that the use of words such as force or pressure is strictly metaphorical, and that there is no such force or pressure connected with selection, as there is in discussions in the physical sciences.

22. Where does chance (stochastic processes) enter the process of selection?
The first step in selection, the production of genetic variation, is almost exclusively a chance phenomenon except that the nature of the changes at a given gene locus is strongly constrained. Chance plays an important role even at the second step, the process of the elimination of less fit individuals. Chance may be particularly important in the haphazard survival during periods of mass extinction.

23. Does selection produce perfection?
Darwin already remarked that selection never produces perfection, but only provides adaptation to existing conditions. For instance, animals and plants

in New Zealand had been selected to be adapted to each other. When English animals and plants were introduced to New Zealand, many of the native species, not being "perfect," that is, not being adapted to the invaders, became extinct. The human species is highly successful even though it has not yet completed the transition from quadrupedal to bipedal life in all of its structures. In that sense it is not perfect.

24. How did human consciousness evolve?

This is a question that psychologists love to ask. The answer is actually quite simple: from animal consciousness! There is no justification in the widespread assumption that consciousness is a unique human property. Students of animal behavior have brought together a great deal of evidence showing how widespread consciousness is among animals. Every dog owner has had occasion to observe the "guilt feeling" a dog displays when, in the absence of its master, he has done something for which he expects to be punished. How far "down" in the animal kingdom one can trace such signs of consciousness is arguable. It may well be involved even in the avoidance reaction of some invertebrates and even protozoans. However, it is quite certain that human consciousness did not arise full-fledged with the human species, but is only the most highly evolved end point of a long evolutionary history.

Acoelomate   An animal that lacks a coelom. The Platyhelminthes are acoelomates.

Adaptation   Any property of an organism believed to add to its fitness.

Adaptationist program   The investigation of the possible adaptive value of a structure or other attribute of a taxon.

Adaptive radiation   Evolutionary divergence of members of a single phyletic line into different niches or adaptive zones.

Allele   One of the alternate forms (nucleotide sequences) of a gene. Different alleles of the same gene usually produce different effects on the phenotype.

Allopatric   Pertaining to populations or species the ranges of which do not overlap.

Allospecies   Species that are members of a superspecies but that are geographically separated from the other allospecies of this superspecies.

Allozyme   The particular amino acid sequence of an enzyme produced by one allele of a gene that also has other alleles producing enzymes with different amino acid sequences.

Alvarez event   The impact of an asteroid on the Earth at the very end of the Cretaceous, 65 million years ago, causing the mass extinction of the dinosaurs and other fauna and flora, as postulated by the physicist Walter Alvarez.

Anagenesis   So-called progressive ("upward") evolution.

Anlage   In development, the propensity of a tissue to give rise to a particular structure or organ.

Anoxia   Deficiency or absence of oxygen.

Anthropomorphism   An unjustified attribution of a human characteristic to other organisms or objects.

Australopithecines Early African hominids, living about 4.4 to 2.0 million years ago, who had a small brain (less than 500 cc), were bipedal, but were still largely arboreally living; they had no stone tools.

Background extinction The steady extinction at all geological periods of a certain number of individual species.

Baldwin effect The selection of genes that strengthen the genetic basis of a variant of the phenotype.

Bauplan **(body plan)** Structural type, as that of a vertebrate or arthropod.

Biological species Groups of actually or potentially interbreeding natural populations that are reproductively isolated from other such groups.

Biota The combined fauna and flora of an area.

Budding The origin of a new side branch of a phyletic lineage by speciation and subsequent entry of this species and its descendants into a new niche or adaptive zone, resulting in a distinct new higher taxon.

Category A taxonomic category designates the rank of a taxon in a hierarchy of levels; a class whose members are all taxa assigned the same categorical rank.

Causation, proximate Causation due to currently acting biological, chemical, or physical factors.

Chromosomes Structural elements, usually rod-shaped, found in the nucleus of a cell and containing the major part of the hereditary material (the genes). Chromosomes are composed of DNA and proteins.

Clade Portion of a phylogenetic tree between two branching points or from a branching point to the end of the branch.

Cladogenesis The branching (divergence) component of evolution.

Cleavage One of the series of mitotic divisions of the fertilized egg (zygote) giving rise to the early embryonic tissues.

Cline Gradual variation of a character in a species, usually parallel to the variation of a climatic or other environmental gradient.

Clone Genetically identical individuals produced by any process of asexual (uniparental) reproduction; also monozygotic twins.

Coalescence method A method based on molecular clock determined rates of divergence to infer the time of the split of two related taxa from the lineage of their common ancestor.

Codon A nucleotide triplet in the genetic program (genome), designating a particular amino acid.

Coevolution The parallel evolution of two kinds of organisms that are interdependent, like flowers and their pollinators, or where at least one depends on the other, like predators on prey or parasites on their hosts, and where any change in one will result in an adaptive response in the other.

Competitive exclusion principle Two species cannot exist at the same locality if they have identical ecological requirements.

Continental drift The movement of continents in geological time owing to the drift of the plates of the Earth's mantle caused by plate tectonics.

Contingency A nonpredictable occurrence.

Convergence Phenotypic similarity of two taxa that is independently acquired and is not produced by a genotype inherited from a common ancestor.

Copying error Failure of a gene to replicate itself precisely during mitosis or meiosis, resulting in a mutation.

Creationism Belief in the literal truth of Creation as recorded in the Book of Genesis.

Crossing-over The exchange of corresponding segments between maternal and paternal chromosomes. It occurs when maternal and paternal homologous chromosomes are paired during prophase of the first meiotic division.

Cynodonts An extinct group of reptiles, ancestral to the mammals.

Daphnia A planktonic crustacean of the order Cladocera.

Darwinism Darwin's concepts and theories on which his followers base the explanation of evolution.

Darwinism, social A political theory postulating that ruthless egotism is the most successful policy.

Deme A local population of potentially interbreeding individuals.

Dendrogram A diagram in the form of a branching tree designed to indicate degrees of relationship among taxa.

Diploid Possessing a double set of chromosomes, one set derived from the mother, the other from the father.

Discontinuity, phenetic A discontinuity (gap) in the range of variation of the phenotypes in a population.

Discontinuity, taxic A discontinuity (gap) in the range of variation among related taxa, such as species of a genus or genera of a family.

Dispersal The movement of individuals from their birthplace; more broadly, the spread of individuals of a species beyond the current species range.

Ecological role The contribution made by a characteristic of an organism to its survival.

Elimination, nonrandom The elimination of the less fit individuals of a population during the process of so-called natural selection.

Entropy The degradation of matter and energy in the universe to an ultimate state of inert uniformity. Entropy can be reached only in a closed system.

Epistasis   Interactions between two or more genes.

Essentialism   A belief that the variation of nature can be reduced to a limited number of basic classes, representing constant, sharply delimited types; typological thinking.

Evolution   The gradual process by which the living world has been developing following the origin of life.

Evolutionary synthesis   The achievement of consensus among previously feuding schools of evolutionists, such as experimental geneticists, naturalists, and paleontologists, taking place particularly in the period 1937–1947; the unification of various branches of evolutionary biology, such as those studying anagenesis and those studying cladogenesis.

Fauna   The species of animals living in a given geographical area at a given time.

Fertilization   Fusion between the male gamete (spermatozoon) and the female gamete (ovum). It results in the joining of a haploid set of maternal chromosomes with a haploid set of paternal chromosomes in the newly formed zygote, which thereby becomes diploid.

Finalism   Belief in an inherent trend in the natural world toward some preordained final goal or purpose, such as the attainment of perfection. See *teleology*.

Flora   The species of plants living in a given geographical area at a given time.

Founder population   A population beyond the previous species range founded by a single female (or a small number of conspecifics).

Gaia hypothesis   The hypothesis that the interactions, particularly chemical ones, between organisms and the inorganic world in which they live (including the atmosphere) are regulated by a control program, called Gaia.

Gamete   A male or female reproductive cell; spermatozoon or ovum (egg).

Gene   A genetic unit (set of base pairs) situated on a particular locus on a chromosome.

Gene flow   The movement of genes in a species from population to population.

Genetic drift   The occurrence of changes in gene frequency brought about not by selection but by chance. It occurs especially in small populations.

Genetic program   The information coded in an organism's DNA.

Genotype   The set of genes of an individual.

Group selection, theory of   The theory that a social group can be the object of selection if the cooperative interaction among the members of the group enhances the fitness of the group.

Haploid   Possessing a single set of chromosomes, like the gametes.

Heliocentricity   The theory that the sun is in the center of the solar system and that the planets circle around the sun.

Heterozygous   Possessing two different alleles of a particular gene on a pair of homologous chromosomes.

Homeostasis, genetic   The capacity of the genotype to compensate for disturbing environmental influences.

Homologous   Referring to the structure, behavior, or other character of two taxa that is derived from the same or equivalent feature of their nearest common ancestor.

Homoplasy   Similarity of characters in two taxa not due to derivation from the same characters in the nearest common ancestor. See *parallelophyly* and *convergence.*

Homozygous   Possessing identical alleles of a particular gene or a pair of homologous chromosomes.

Infusorian   Obsolete term for small aquatic organisms (mostly protozoans, crustaceans, rotifers, and one-celled algae).

Isolating mechanism   Genetic (including behavioral) properties of individuals that prevent populations of different species from interbreeding where they coexist in the same area.

Kin selection   Selective advantage due to the altruistic interaction of individuals sharing part of the same genotype, such as siblings.

Linnaean   Named for the Swedish naturalist Carolus Linnaeus (1707–1778), who invented the binomial classification system.

Living fossil   A living species surviving after all of its relatives have become extinct more than 50–100 million years ago.

Locus   The position of a particular gene on a chromosome.

Macroevolution   Evolution above the species level; the evolution of higher taxa and the production of evolutionary novelties, such as new structures.

Mass extinction   The extermination of a large proportion of the biota on Earth by a climatic, geological, cosmic, or other environmental event.

Meiosis   A special form of nuclear division that occurs during the formation of the gametes (spermatozoa and eggs) in sexually reproducing organisms. Crossing-over and the reduction division of the chromosomes take place during meiosis.

Microevolution   Evolution at or below the species level.

Mimicry, Batesian   Resemblance of a palatable species to an unpalatable or toxic one.

Mimicry, Müllerian   Resemblance of an unpalatable or toxic species to another likewise unpalatable one.

Missing link   A fossil bridging the large gap between an ancestral and a derived group of organisms, such as *Archaeopteryx*, between reptiles and birds.

Mitosis   A form of cell division in which each chromosome "splits" lengthwise (it replicates itself), each daughter cell receiving one daughter chromosome. This is the typical division of somatic cells.

Molecular clock   The clocklike regularity of the change of a molecule (gene) or a whole genotype over geological time.

Mosaic evolution   Evolutionary change that occurs in a taxon at different rates for different structures, organs, or other components of the phenotype.

Mutation   Any inheritable alteration in the genetic material, most commonly an error of replication during cell division, resulting in the replacement of an allele by a different one. In addition to such gene mutations, there are also chromosomal mutations, i.e., major chromosomal changes, including polyploidy.

Natural selection   The process by which in every generation individuals of lower fitness are removed from the population.

Necessity   The inevitable force of circumstances.

Niche   A constellation of properties of the environment making it suitable for occupation by a species.

Normalizing (stabilizing) selection   The elimination by selection of variants beyond the normal range of variation of a population.

Open reading frame   DNA sequence that potentially can be translated into a protein.

Organizer   A tissue capable of inducing a specific type of development in other undifferentiated tissues.

Orthogenesis   The refuted hypothesis that rectilinear trends in evolution are caused by an intrinsic finalistic principle.

Orthologous genes   Genes in different species that are sufficiently similar in their nucleotide sequences to indicate that they were derived from a common ancestor.

Panmictic   Pertaining to populations and species of such great dispersal capacity that there is complete interbreeding of populations from all parts of their range.

Parallelophyly   Multiple independent occurrence of the same character in different species derived from the nearest common ancestor that has the genetic disposition for this character but did not show it in its own phenotype.

Parapatric   Pertaining to contiguously living but nonoverlapping populations or species.

Phenotype The total of all observable features of a developing or developed individual (including its anatomical, physiological , biochemical, and behavioral characteristics). The phenotype is the result of interaction between the genotype and the environment.

Philopatry The drive (tendency) of an individual to return to (or stay in) its home area (birthplace or another adopted area).

Phyletic evolution The evolutionary change of a phyletic lineage in the time dimension.

Phyletic lineage A branch of the phylogenetic tree; all the linear descendants of an ancestral species.

Phylogeny The inferred lines of descent of a group of organisms, including a reconstruction of the common ancestor and the amount of divergence of the various branches.

Plate A piece of the Earth's crust that moves owing to plate tectonics.

Plate tectonics The theory that the crust of the Earth consists of movable plates that may join or separate in different geological periods.

Pleiotropic Pertaining to how a gene may affect several aspects of the phenotype.

Polygenic inheritance Inheritance of a trait (e.g., height) governed by several genes (polygenes or multiple factors). Their effect is cumulative.

Polymorphism The simultaneous occurrence of several different alleles or discontinuous phenotypes in a population, with the frequency of even the rarest type higher than can be maintained by recurrent mutation.

Polymorphism, balanced The condition in which two different alleles coexisting in the same population produce a heterozygote of greater fitness than either homozygote.

Polyphyly Derivation of a taxon from two or more different ancestral sources.

Preadapted Pertaining to a character capable of adopting a new function or ecological role without loss of fitness; the possession of the required properties to permit a shift into a new niche or habitat, without interference with the original functions.

Protists A convenient collective name for the vast variety of unicellular eukaryotes.

Punctuated equilibria Alternation of extremely rapid and normal or slow evolutionary change in a phyletic lineage, as a result of speciational evolution.

Recapitulation The appearance of a structure or other attribute of a larval or immature individual of a species that resembles a similar attribute of the adults of an ancestral species; it is interpreted as evidence for descent from that ancestor.

Recessive gene   A gene that is unable to express its effect when it is present in the heterozygous state (single dose). It must be present in the homozygous state (double dose) to express its effect.

Recombination   A reshuffling of the genes in a new zygote as a result of crossing-over and reassortment of the chromosomes during meiosis. A new set of genotypes is thus produced in each generation.

Reductionism   The belief that the higher levels of integration of a complex system can be fully explained through a knowledge of the smallest components.

Saltation   A sudden event, resulting in a discontinuity (gap), such as the sudden production of a new species or higher taxon.

Saltationism   The belief that evolutionary change is the result of the sudden origin of a new kind of individual that becomes the progenitor of a new kind of organism.

Scala naturae   A linear arrangement of all forms of life from the lowest, nearly inanimate to the most perfect; the Great Chain of Being.

Scientific Revolution   The period in the sixteenth and seventeenth centuries in which scientists, including Galileo and Newton, laid the foundation of modern science.

Sex-linkage   The type of linkage produced when a gene is located on the X or the Y chromosome.

Sexual selection   Selection for attributes that enhance reproductive success.

Sickle cell disease   A genetic disease of the red blood corpuscles. Homozygosity for the sickle cell gene results in early death, while heterozygotes have superior fitness in malarial regions.

Somatic mutation   The occurrence of a mutation in a somatic cell.

Somatic program   In development, the information contained in neighboring tissues that may influence or control the further development of an embryonic structure or tissue.

Speciation, allopatric   The origin of a new species through the acquisition of effective isolating mechanisms by a geographically isolated portion of the parental species.

Speciation, dichopatric   The origin of a new species through the division of a parental species by a geographical, vegetational, or other extrinsic barrier.

Speciation, peripatric   The origin of new species through the modification of peripherally isolated founder populations. See *budding*.

Speciation, sympatric   Speciation without geographical isolation; the origin of a new set of isolating mechanisms within a deme.

Speciational evolution   Accelerated evolutionary change toward species status in a founder or relict population, sometimes leading to the origin of a new higher taxon.

Species concept   The biological meaning or definition of the word "species"; the criteria on the basis of which a species taxon is delimited.

Species taxon   A taxon qualifying as a species according to the accepted species concept.

Spontaneous generation   A refuted early concept that complex organisms can be produced spontaneously from inanimate material.

Stasis   A period in the history of a taxon during which evolution seemed to have been at a standstill.

Symbiosis   The usually mutually beneficial interaction of individuals of two different species.

Sympatric   Pertaining to species the ranges of which overlap; species coexisting in the same area.

Taxon   A monophyletic group of organisms (or lower taxa) that can be recognized by sharing a definite set of characters.

Teleology   The study of final causes; the belief in the existence of direction-giving forces.

Therapsida   An order of fossil synapsid reptiles that gave rise to the mammals.

Transformationism   The refuted theories that attributed evolution to a change of the essence of a species either by inheritance of acquired characters, or by direct influence of the environment, or by final causes.

Transmutationism   The theory that evolutionary change is caused by sudden new mutations or saltations, producing instantaneously a new species. See *saltationism*.

Typological species concept   The recognition of species on the basis of their degree of phenotypic difference.

Typologist   One who disregards variation and considers the members of a population to be replicas of the type; an essentialist.

Uniformitarianism   The theory of some pre-Darwinian geologists, particularly Charles Lyell, that all changes in the Earth's history are gradual, rather than occurring in saltations or jumps. Being gradual, these changes cannot be considered acts of special creation.

Vestigial character   A deconstructed, nonfunctional characteristic that had been fully functional in a species' ancestor, like the eyes in cave animals and the human appendix.

Wallace's Line   In biogeography, a line through the Indo-Malayan archipelago that indicates the eastern edge of the continental Sunda Shelf, serving as the eastern limit of the range of much of the tropical Asian mainland fauna, particularly in mammals.

Zygote   A fertilized egg; the individual that results from the union of two gametes and their nuclei.

Alters, B. J., and S. M. Alters. 2001. *Defending Evolution in the Classroom*. Sudbury, Mass.: Jones and Bartlett.

Anderson, M. 1994. *Sexual Selection*. Princeton: Princeton University Press.

Arnold, Michael L. 1997. *Natural Hybridization and Evolution*. Oxford: Oxford University Press.

Avery, O. T., C. M. MacLeod, and M. McCarthy. 1944. Studies on the chemical nature of the substance inducing transformation of pneumococcal types. I. Induction of transformation by a deoxyribonucleic acid fraction isolated from pneumococcus type III. *Journal of Experimental Medicine* 79: 137–158.

Avise, John. 2000. *Phylogeography*. Cambridge, Mass.: Harvard University Press.

Baer, K. E. von. 1828. *Entwicklungsgeschichte der Thiere*. Königsberg: Bornträger.

Bartolomaeus, T. 1997/1998. Chaetogenesis in polychaetous Annelida. *Zoology* 100: 348–364.

Bates, H. W. 1862. Contributions to an insect fauna of the Amazon Valley. *Trans. Linn. Soc. London* 23: 495–566.

Bekoff, M. 2000. Animal emotions: Exploring passionate natures. *Bioscience* 50: 861–870.

Bell, G. 1996. *Selection. The Mechanisms of Evolution*. New York: Chapman and Hall.

Berra, Tim M. 1990. *Evolution and the Myth of Creationism*. Stanford: Stanford University Press.

Bock, G. R., and G. Cardew (eds.). 1999. *Homology. Novartis Symposium*. New York: John Wiley & Sons.

Bodmer, W., and R. McKie. 1995. *The Book of Man: The Quest to Discover Our Genetic Heritage*. London: Abacus.

Bonner, J. T. 1998. The origins of multicellularity. *Integrative Biology*, pp. 27–36.

Bowler, Peter J. 1996. *Life's Splendid Drama: Evolutionary Biology and the Reconstruction of Life's Ancestry*. Chicago: University of Chicago Press.

Brack, André (ed.). 1999. *The Molecular Origins of Life: Assembling Pieces of the Puzzle*. Cambridge: Cambridge University Press.

Brandon, R. N. 1995. *Concepts and Methods in Evolutionary Biology*. Cambridge: Cambridge University Press.

Bush, G. L. 1994. Sympatric speciation in animals. *TREE* 9: 285–288.

Butler, A. B., and W. M. Saidel. 2000. Defining sameness: Historical, biological, and generative homology. *Bioessays* 22: 846–853.

Cain, A. J., and P. M. Sheppard. 1954. Natural selection in *Cepaea*. *Genetics* 39: 89–116.

Campbell, Neil A., et al. 1999. *Biology*, 5th ed. Menlo Park, Calif.: Benjamin Cummings.

Cavalier-Smith, T. 1998. A revised six-kingdom system of life. *Biol. Rev.* 73: 203–266.

Chatterjee, Sankar. 1997. *The Rise of Birds: 225 Million Years of Evolution*. Baltimore: Johns Hopkins University Press.

Cheetham, A. H. 1987. Tempo in evolution in a neogene bryozoan. *Paleobiology* 13: 286–296.

Corliss, J. O. 1998. Classification of protozoa and protists: The current status. In G. H. Coombs, K. Vickerman, M. A. Sleigh, and A. Warren (eds.), *Evolutionary Relationships Among Protozoa*, pp. 409–447. London: Chapman and Hall.

Cracraft, Joel. 1984. The terminology of allopatric speciation. *Syst. Zool.* 33: 115–116.

Cronin, H. 1991. *The Ant and the Peacock*. Cambridge: Cambridge University Press.

Cuvier, G. 1812. *Recherches sur les ossemens fossiles des quadrupèdes,....* 4 vols. Paris: Déterville.

Darwin, C. 1859. *On the Origin of Species*. London: John Murray.

———. 1871. *The Descent of Man*. London: John Murray.

Dawkins, Richard. 1982. *The Extended Phenotype: The Gene as the Unit of Selection*. Oxford: Freeman.

———. 1986. *The Blind Watchmaker*. New York: W. W. Norton.

———. 1995. *River Out of Eden: A Darwinian View of Life*. New York: Basic Books.

———. 1996. *Climbing Mount Improbable*. New York: W. W. Norton.

de Waal, Frans. 1997. *Good Natured: The Origin of Right and Wrong in Human and Other Animals.* Cambridge, Mass.: Harvard University Press.

Dobzhansky, R., and O. Pavlovsky. 1957. An experimental study of interaction between genetic drift and natural selection. *Evolution* 11: 311–319.

Ehrlich, P. 2000. *Human Natures.* Washington, D.C.: Island Press.

Ehrlich, P., and D. H. Raven. 1965. Butterflies and plants: A study in coevolution. *Evolution* 18: 586–608.

Eldredge, N. 2000. *The Triumph of Evolution and the Failure of Creationism.* New York: W. H. Freeman.

Eldredge, N., and S. J. Gould. 1972. Punctuated equilibria: An alternative to phyletic gradualism. In T. J. M. Schopf and J. M. Thomas (eds.), *Models in Paleobiology*, pp. 82–115. San Francisco: Freeman, Cooper.

Endler, John A. 1986. *Natural Selection in the Wild.* Princeton: Princeton University Press.

Erwin, D., J. Valentine, and D. Jablonski. 1997. The origin of animal body plans. *American Scientist* 85: 126–137.

Fauchald, K., and G. W. Rouse. 1997. Polychaete systematics: Past and present. *Zool. Scripte.* 26: 71–138.

Feduccia, Alan. 1999. *The Origin and Evolution of Birds,*. 2nd ed. New Haven: Yale University Press.

Freeman, Scott, and Jon C. Herron. 2000. *Evolutionary Analysis.* New York: Prentice Hall.

Futuyma, Douglas J. 1983. *Science on Trial. The Case for Evolution.* New York: Pantheon Books.

_____. 1998. *Evolutionary Biology*, 3rd ed. Sunderland, Mass.: Sinauer Associates.

Gehring, W. J. 1999. *Master Control Genes in Development and Evolution.* New Haven: Yale University Press.

Geoffroy, St. Hilaire, Étienne. 1822. *La Loi de Balancement.* Paris.

Gesteland, R., T. Cech, and J. Atkins. 1999. *The RNA World.* Cold Spring Harbor Laboratory Press.

Ghiselin, Michael T. 1996. Charles Darwin, Fritz Müller, Anton Dohrn, and the origin of evolutionary physiological anatomy. *Memorie della Societa Italiana di Scienze Naturali e del Museo Civico di Storia Naturale di Milano* 27: 49–58.

Giribet, G., D. L. Distel, M. Polz, W. Sterner, and W. C. Wheeler. 2000. Triploblastic relationships with emphasis on the acoelomates and the position of Gnathostomulida, Cycliophora, Plathelminthes, and Chaetognatha. *Syst. Biol.* 49: 539–562.

Givnish, T. J., and K. J. Sytsma (eds.). 1997. *Molecular Evolution and Adaptive Radiation*. Cambridge: Cambridge University Press.

Goldschmidt, R. 1940. *The Material Basis of Evolution*. New Haven: Yale University Press.

Gould, S. J. 1977. The return of hopeful monsters. *Natural History* 86 (June/July): 22–30.

_____. 1989. *Wonderful Life: The Burgess Shale and the Nature of History*. New York: W. W. Norton.

Gould, S. J., and R. Lewontin. 1979. The spandrels of San Marco and the Panglossian paradigm: A critique of the adaptationist programme. *Proceedings of the Royal Society of London, Series B* 205: 581–598.

Gram, D., and W. H. Li. 1999. *Fundamentals of Molecular Evolution*, 2nd ed. Sunderland, Mass.: Sinauer Associates.

Grant, Verne. 1963. *The Origin of Adaptations*. New York: Columbia University Press.

_____. 1981. *Plant Speciation*, 2nd ed. New York: Columbia University Press.

_____. 1985. *The Evolutionary Process*. New York: Columbia University Press.

Graur, Dan, and Wen-Hsiung Li. 1999. *Fundamentals of Molecular Evolution*, 2nd ed. Sunderland, Mass.: Sinauer Associates.

Gray, Asa. 1963 [1876]. *Darwiniana* (new edition, A. H. Dupree, ed.), pp. 181–186. Cambridge, Mass.: Harvard University Press.

Griffin, Donald R. 1981. *The Question of Animal Awareness: Evolutionary Continuity of Mental Experience*, rev. ed. Los Altos, Calif.: Kaufmann.

_____. 1984. *Animal Thinking*. Cambridge, Mass.: Harvard University Press.

_____. 1992. *Animal Minds*. Chicago: University of Chicago Press.

Haeckel, E. 1866. *Generelle Morphologie der Organismen*. Berlin: Georg Reimer.

Haldane, J. B. S. 1929. The origin of life. *Rationalist Ann.*, p. 3.

_____. 1932. *The Causes of Evolution*. New York: Longman, Green.

Hall, B. K. 1998. *Evolutionary Developmental Biology*, 2nd ed. Norwell, Mass.: Kluwer Academic Publishers.

_____. 2001. *Phylogenetic Trees Made Easy*. Sunderland, Mass.: Sinauer Associates.

Hamilton, W. D. 1964. The genetic evolution of social behavior. *J. Theoretical Biology* 7: 1–52.

Hartl, Daniel L., and Elizabeth W. Jones. 1999. *Essential Genetics*, 2nd ed. Sudbury, Mass.: Jones and Bartlett.

Hatfield, T., and D. Schluter. 1999. Ecological speciation in sticklebacks: Environment dependent fitness. *Evolution* 53: 866–879.

Hines, P., and E. Culotta. 1998. The evolution of sex. *Science* 281: 1979–2008.

Hopson, J. A., and H. R. Barghusen. 1986. An analysis of therapsid relationships. In N. Hotton III et al. (eds.), *The Ecology and Biology of Mammal-like Reptiles*, pp. 83–106. Washington/London: Smithsonian Institution Press.

Howard, D. J., and S. H. Berlocher (eds.). 1998. *Endless Forms: Species and Speciation*. New York: Oxford University Press.

Huxley, T. H. 1863. *Evidence as to Man's Place in Nature*.

_____. 1868. On the animals which are most closely intermediate between the birds and the reptiles. *Ann. Mag. Nat. Hist.* 2: 66–75.

Jacob, F. 1977. Evolution and tinkering. *Science* 196: 1161–1166.

Kay, Lily E. 2000. *Who Wrote the Book of Life? A History of the Genetic Code*. Stanford: Stanford University Press.

Keller, E. F., and E. A. Lloyd. 1992. *Keywords in Evolutionary Biology*. Cambridge, Mass.: Harvard University Press.

Keller, L. (ed.). 1999. *Levels of Selection in Evolution*. Princeton: Princeton University Press.

Kimura, Motoo. 1983. *The Neutral Theory of Molecular Evolution*. Cambridge: Cambridge University Press.

Kirschner, M., and J. Gerhart. 1998. Evolvability. *Proceedings of the National Academy of Sciences* 98: 8420–8427.

Kitcher, Philip. 1982. *Abusing Science. The Case Against Creationism*. Cambridge, Mass.: MIT Press.

Lack, David. 1947. *Darwin's Finches*. Cambridge: Cambridge University Press.

Lamarck, Jean-Baptiste. 1809. *Philosophie Zoologique*. Paris.

Lawrence, P. A. 1992. *The Making of a Fly*. London: Blackwell.

Li, W. H. 1997. *Molecular Evolution*. Sunderland, Mass.: Sinauer Associates.

Lovejoy, A. B. 1936. *The Great Chain of Being*. Cambridge, Mass.: Harvard University Press.

Magurran, Ann E., and Robert M. May (eds.). 1999. *Evolution of Biological Diversity*. Oxford/New York: Oxford University Press.

Margulis, L. 1981. *Symbiosis in Cell Evolution*. San Francisco: W. H. Freeman.

_____. 1996. Archaeal–eubacterial mergers in the origin of Eukarya. Phylogenetic classification of life. *Proceedings of the National Academy of Sciences* 93: 1071–1076.

Margulis, Lynn, and Rene Fester (eds.). 1991. *Symbiosis as a Source of Evolutionary Innovation*. Cambridge, Mass.: MIT Press.

Margulis, L., and K. V. Schwartz. 1998. *Five Kingdoms*, 3rd ed. New York: W. H. Freeman.

Margulis, Lynn, Dorion Sagan, and Lewis Thomas. 1997. *Microcosmos: Four Billion Years of Evolution from Our Microbial Ancestors*. Berkeley: University of California Press.

Margulis, Lynn, Michael F. Dolan, and Ricardo Guerrero. 2000. The chimeric eukaryote: Origin of the nucleus from the karyomastigont in amitochondriate protists. *Proceedings of the National Academy of Sciences* 97: 6954–6959.

Marshall, Charles, and J. W. Schopf (eds.). 1996. *Evolution and the Molecular Revolution*. Sudbury, Mass.: Jones and Bartlett.

Martin, W., and M. Müller. 1998. The hydrogen hypothesis for the first eukaryote. *Nature* 392: 37–41.

Masson, V. J., and Susan McCarthy. 1995. *When Elephants Weep: The Emotional Lives of Animals*. New York: Delacorte Press.

May, R. 1990. How many species? *Philos. Trans. Roy. Soc. London, Ser. B* 330: 293–301; (1994) 345: 13–20.

———. 1998. The dimensions of life on earth. In *Nature and Human Society*. Washington, D.C.: National Academy of Sciences.

Maynard Smith, J. 1982. *Evolution and the Theory of Games*. Cambridge: Cambridge University Press.

———. 1989. *Evolutionary Genetics*. Oxford: Oxford University Press.

Maynard Smith, J., and E. Szathmary. 1995. *The Major Transitions in Evolution*. Oxford: Freeman/Spektrum.

Mayr, Ernst. 1942. *Systematics and the Origin of Species*. New York: Columbia University Press.

———. 1944. Wallace's line in the light of recent zoogeographic studies. *Quarterly Review of Biology* 19: 1–14.

———. 1954. Change of genetic environment and evolution. In J. Huxley, A. C. Hardy, and E. B. Ford (eds.), *Evolution as a Process*, pp. 157–180. London: Allen and Unwin.

———. 1959. Darwin and the evolutionary theory in biology. In *Evolution and Anthropology: A Centennial Appraisal*, pp. 1–10. Washington, D.C.: Anthropological Society of America.

———. 1960. The emergence of evolutionary novelties. In Sol Tax (ed.), *Evolution after Darwin. I. The Evolution of Life*, pp. 349–380. Chicago: University of Chicago Press.

———. 1963. *Animal Species and Evolution*. Cambridge, Mass.: Harvard University Press.

———. 1969. *Principles of Systematic Zoology*. New York: McGraw-Hill.

_____. 1974. Behavior programs and evolutionary strategies. *American Scientist* 62: 650–659.

_____. 1982. *The Growth of Biological Thought: Diversity, Evolution, and Inheritance.* Cambridge, Mass.: Harvard University Press.

_____. 1983. How to carry out the adaptationist program? *American Naturalist* 121: 324–334.

_____. 1986. The philosopher and the biologist. Review of *The Nature of Selection: Evolutionary Theory in Philosophical Focus* by Elliott Sober (MIT Press, 1984). *Paleobiology* 12: 233–239.

_____. 1991. *Principles of Systematic Zoology,* rev. ed. with Peter Ashlock. New York: McGraw-Hill.

_____. 1992. Darwin's principle of divergence. *J. Hist. Biol.* 25: 343–359.

_____. 1994. Recapitulation reinterpreted: The somatic program. *Onart. Rev. Biol.* 64: 223-232.

_____. 1997. The objects of selection. *Proceedings of the National Academy of Sciences* 94: 2091–2094.

Mayr, Ernst, and J. Diamond. 2001. *The Birds of Northern Melanesia.* New York: Oxford University Press.

Mayr, Ernst, and W. Provine (eds.). 1980. *The Evolutionary Synthesis* (2nd ed. with new foreword published in 1999). Cambridge, Mass.: Harvard University Press.

McHugh, D. 1997. Molecular evidence that echiurans and pogonophorans are derived annelids. *Proceedings of the National Academy of Sciences* 94: 8006–8009.

Michod, Richard E., and Bruce R. Levin. 1988. *The Evolution of Sex.* Sunderland, Mass.: Sinauer Associates.

Midgley, M. 1994. *The Ethical Primate.* London: Routledge.

Milkman, R. 1982. *Perspectives on Evolution.* Sunderland, Mass.: Sinauer Associates.

Montagu, Ashley (ed.). 1983. *Science and Creationism.* New York: Oxford University Press.

Moore, J. A. 2001. *From Genesis to Genetics.* Berkeley: University of California Press.

Morgan, T. H. 1910. Chromosomes and heredity. *American Naturalist* 44: 449–496.

Morris, S. Conway. 2000. The Cambrian "explosion": Slow fuse or megatonnage? *Proceedings of the National Academy of Sciences* 97: 4426–4429.

Muller, Fritz. 1864. *Für Darwin.* In A. Moller (ed.), *Fritz Müller, Werke, Briefe, und Leben.* Jena: Gustao Fischer.

Nevo, Eviatar. 1995. Evolution and extinction. In W. A. Nierenberg (ed.),

Encyclopedia of Environmental Biology, vol. 1, pp. 717–745. San Diego, Calif.: Academic Press.

_____. 1999. *Mosaic Evolution of Subterranean Mammals: Regression, Progression, and Global Convergence.* New York: Oxford University Press.

Newell, Norman D. 1982. *Creation and Evolution: Myth or Reality*: New York: Columbia University Press.

Nitecki, Matthew H. (ed.). 1984. *Extinctions.* Chicago: University of Chicago Press.

_____. 1988. *Evolutionary Progress.* Chicago: University of Chicago Press.

Oparin, A. I. 1938. *The Origin of Life.* New York: Macmillan.

Page, R. D. M., and E. C. Holmes. 1998. *Molecular Evolution: A Phylogenetic Approach.* Oxford: Blackwell Science.

Paley, William. 1802. *Natural Theology: On Evidences of the Existence and the Attributes of the Deity.* London: R. Fauldner.

Paterson, Hugh E. H. 1985. The recognition concept of species. In E. S. Verba (ed.), *Species and Speciation*, Transvaal Museum Monograph No. 4, pp. 21–29. Pretoria, South Africa: Transvaal Museum.

Peacocke, A. R. 1979. *Creation and the World of Science.* Oxford: Clarendon Press.

Pickford, M., and B. Senut. 2001. *Comptes Rend. Acad. Sci.*

Raff, R. A. 1996. *The Shape of Life. Development and the Evolution of Animal Form.* Chicago: University of Chicago Press.

Ray, John. 1691. *The Wisdom of God Manifested in the Works of the Creator.*

Rensch, B. 1947. *Neuere Probleme der Abstammungslehre.* Stuttgart: Enke.

Rice, W. R. 1987. Speciation via habitat specialization: The evolution of reproductive isolation as correlated character. *Evolution and Ecology* 1: 301–314.

Ridley, Mark. 1996. *Evolution*, 2nd ed. Cambridge, Mass.: Blackwell Science.

Riesenberg, Loren H. 1997. Hybrid origins of plant species. *Annual Review of Ecology and Systematics* 28: 359–389.

Ristan, Carolyn A. (ed.). 1991. *Cognitive Ethology: The Minds of Other Animals.* Hillsdale, N.J.: Lawrence Erlbaum Associates.

Rizzotti, M. 1996. *Defining Life.* Padova: University of Padova.

_____. 2000. *Early Evolution: From the Appearance of the First Cell to the First Modern Organisms.* Boston: Birkhäuser.

Rose, Michael R., and G. V. Lander (eds.). 1996. *Adaptation.* San Diego, Calif.: Academic Press.

Rüber, L., E. Verheyen, and Axel Meyer. 1999. Replicated evolution of trophic specializations in an endemic cichlid fish lineage from Lake Tanganyika. *Proceedings of the National Academy of Sciences* 96: 10,230–10,235.

Ruse, Michael. 1982. *Darwinism Defended*. Reading, Mass.: Addison & Wesley.
_____. 1998 [1986]. *Taking Darwin Seriously*. Amherst, N.Y.: Prometheus Books.

Sagan, Dorion, and Lynn Margulis. 2001. Origin of eukaryotes. In S. A. Levin (ed.), *Encyclopedia of Biodiversity*, Vol. 2, pp. 623–633. San Diego, Calif.: Academic Press.

Salvini Plawen, L., and Ernst Mayr. 1977. On the evolution of photoreceptors and eyes. *Evolutionary Biology* 10: 207–263.

Sanderson, Michael, and Larry Hufford (eds.). 1996. *Homoplasy: The Recurrence of Similarity in Evolution*. San Diego, Calif.: Academic Press.

Sapp, J. 1994. *Evolution by Association: A History of Symbiosis*. New York/Oxford: Oxford University Press.

Schindewolf, H. O. 1950. *Grundfragen der Paläontologie*. Stuttgart: Schweizerbart.

Schopf, J. W. 1999. *Cradle of Life*. Princeton: Princeton University Press.

Simpson, G. G. 1953. *The Major Features of Evolution*. New York: Columbia University Press.

Singh, R. S., and C. B. Krimbas (eds.). 2000. *Evolutionary Genetics: From Molecules to Morphology*. Cambridge/New York: Cambridge University Press.

Sober, E., and D. S. Wilson. 1998. *Unto Others*. Cambridge, Mass.: Harvard University Press.

Stanley, Steven M. 1998. *Children of the Ice Age: How a Global Catastrophe Allowed Humans to Evolve*. New York: W.H. Freeman.

Starr, Cecie, and Ralph Taggart. 1992. *Diversity of Life*. Pacific Grove, Calif.: Brooks/Cole.

Stewart, W. N. 1983. *Paleobotany and the Evolution of Plants*. Cambridge: Cambridge University Press.

Strait, D. S., and B. A. Wood. 1999. Early hominid biogeography. *Proceedings of the National Academy of Sciences* 96: 9196–9200.

Strickberger, Monroe W. 1985. *Genetics*, 3rd ed. New York: Prentice Hall.
_____. 1996. *Evolution*, 2nd ed. Sudbury, Mass.: Jones and Bartlett.

Sussmen, Robert. 1997. *Biological Basis of Human Behavior*. New York: Simon and Schuster Custom Publishing.

Tattersall, I., and J. H. Schwartz. 2000. *Extinct Humans*. New York: Westview Press.

Taylor, T., and E. Taylor. 1993. *The Biology and Evolution of Fossil Plants*. New York: Prentice Hall.

Thompson, J. N. 1994. *The Coevolutionary Process*. Chicago: University of Chicago Press.

Vanosi, S. M., and D. Schluter. 1999. Sexual selection against hybrids between sympatric stickleback species. Evidence from a field experiment. *Evolution* 53: 874–879.

Vernadsky, Vladimir I. 1926 [1998]. *Biosfera [The Biosphere]*. Forward by Lynn Margulis et al.; introduction by Jacques Grinevald; translated by David B. Langmuir; revised and annotated by Mark A. S. McMenamin. New York: Copernicus.

Wake, D. B. 1997. Incipient species formation in salamanders of the *Ensatina* complex. *Proceedings of the National Academy of Sciences* 94: 7761–7767.

Wakeford, T. 2001. *Liaisons of Life: How the Unassuming Microbe Has Driven Evolution*. New York: John Wiley & Sons.

Watson, James D., and F. Crick. 1953. Molecular structure of nucleic acid. *Nature* 171: 737–738.

West-Eberhard, W. J. 1992. Adaptation. Current usages. In E. F. Keller and E. A. Lloyd (eds.), *Keywords in Evolutionary Biology*, pp. 13–18. Cambridge, Mass.: Harvard University Press.

Westoll, T. Stanley. 1949. On the evolution of the Dipnoi. In Glenn L. Jepsen, Ernst Mayr, and George Gaylord Simpson (eds.), *Genetics, Paleontology, and Evolution*. Princeton: Princeton University Press.

Wheeler, Quentin D., and Rudolf Meier (eds.). 2000. *Species Concepts and Phylogenetic Theory: A Debate*. New York: Columbia University Press.

Willis, J. C. 1940. *The Course of Evolution*. Cambridge: Cambridge University Press.

Wills, C., and Jeffry Bada. 2000. *The Spark of Life*. Boulder: Perseus Books.

Wilson, James Q. 1993. *The Moral Sense*. New York: The Free Press.

Wolf, J. B., E. D. Bradie, and M. J. Wade. 2000. *Epistasis and the Evolutionary Process*. New York: Oxford University Press.

Wrangham, Richard W. 2001. Out of the pan and into the fire: From ape to human. In F. de Waal (ed.), *Tree of Origins*. Cambridge, Mass.: Harvard University Press.

Wright, R. 1994. *The Moral Animal: Evolutionary Psychology and Everyday Life*. New York: Pantheon Books.

Wright, S. 1931. Evolution in Mendelian populations. *Genetics* 16: 97–159.

Young, Willard. 1985. *Fallacies of Creationism*. Calgary, Alberta, Canada: Detrelig Enterprises.

Zahavi, Amotz. 1997. *The Handicap Principle: A Missing Piece of Darwin's Puzzle*. New York/Oxford: Oxford University Press.

Zimmer, Carl. 1998. *At the Water's Edge: Macroevolution and the Transformation of Life*. New York: Free Press.

Zubbay, G. 2000. *Origins of Life on Earth and in the Cosmos.* San Diego, Calif.: Academic Press.

Zuckerkandl, E., and L. Pauling. 1962. In M. Kasha and B. Pullmann (eds.), *Horizons in Biochemistry*, pp. 189–225. New York: Academic Press.

Continental drift, 32, 285
Convergence, 62, 156; definition, 285
Convergent evolution, 222–23(fig.)
Cooking, 245
Cope's Law, 217
Copying error, 285. *See also* Mutation
Correlated evolution, 218–19
Creationism, xiii, 4, 31, 34, 269; definition, 285; natural theology, 148
Cretaceous period, 20(fig.), 25, 59, 234; Alvarez event, 133, 201, 203; bipedal dinosaurs, 227; birds, 65; plants, 63, 64
Crick, F., 112
CroMagnons, 250–51
Crossing-over (meiosis), 103(box), 104, 285, 287
Crustaceans, 162(table); free-swimming larval stages of barnacles, 29(fig.), 30
Cryptic species, 166
Culture: ethics, 259–60; painting, 251; tools, 241, 243, 244, 255. *See also* Religion
Cuvier, Georges, 24–25, 50
Cyanobacteria, 42, 212, 216; mitochondria and, 48; morphological stasis of, 47
Cynodonts, 63, 285

Daphnia, 285
Darwin, Charles, 9–11, 41; branching evolution and, 19; common descent and, 21, 22; fossil record gaps and, 191; on natural selection, 140; Newton and, 76; parapatric speciation and, 182; population thinking and, 75; on struggle, 124, 125; variable populations and, 84–85, 114; variational evolution and, 84–85
Darwinism: macroevolution and, 188–89; major theories, 86(box), 275; major theories, early rejection of, 87(box); molecular biology and, 277; vs. philosophy, 73–74; problems of, 272–73; revolutions of, 86, 275; as a theory of science, 277
Darwinism, social, 285
Demes, 117; definition, 285; phenetic discontinuity and, 192
Dendrograms, 285. *See also* Phylogenetic trees
Deoxyribonucleic acid. *See* DNA
Descent of Man (Darwin), 233
Deuterostomes, 54, 55, 60, 61, 64
Development (organismic), 270; *Archaeopteryx* and, 144; evolutionary

role of, 144; *Hox* genes and, 110, 111(box)
Developmental biology, 268
Devonian period, plants, 63
Dichopatric speciation, 178, 290; vs. peripatric, 179(fig.)
Dimorphism, sexual, 139, 241, 244; *Homo*, 245, 249
Dinosaurs: birds and, 62, 65, 67(fig.), 68(box); extinction of, 201, 276; vs. mammals, 133–34
Diploblasts, 61
Diploids, 90; definition, 285; principles of inheritance and, 91
Discontinuity: phenetic, 192, 285; range, 34, 117; taxic, 192, 285
Disease, 93, 281; European vs. Native Americans, 211; malaria, 122, 126, 195–96; myxomatosis virus, 211; old age and, 142; pathogens, 105, 122, 195, 211; schizophrenia, 127
Dispersal, 32, 98–99, 285; discontinuous distribution, 31; insular distribution, 34; peripheral populations and, 194; range discontinuity, 34, 117; specacular (Brachylophus), 33(fig.)
Diversity, 209–10; among multicellular organisms, 229; of eukaryotes, 228; of phenotypes, 89; of protists, 49; of taxa, 57. *See also* Biodiversity; Variation
DNA (deoxyribonucleic acid), 37, 112, 266, 284, 286; bacteria, 46(box), 47; coalescence method of age determination and, 198(box); coding of various organisms, 38(table); double helix, 92(fig.); functional classes of genes and, 143; mutation and, 96–97; noncoding, 88, 108(box), 109; open reading frame, 288; origin of life and, 43; phylogeny and, 51; principles of inheritance and, 91; transposable elements, 100. *See also* Chromosomes; Double Helix; Nucleic acids
Dobzhansky, T., 39, 89
Drosophila, 109, 193, 194

Earth, the: chronology of animal evolution, 58; plates, 289. *See also* Geological strata; Life, origin of
Ecdysozoa, 56, 61
Ecological role, 285